Hamlyn all-colour paperbacks

John Sparks

Bird Behaviour

illustrated by David Andrews

Paul Hamlyn · London
Sun Books · Melbourne

FOREWORD

Until quite recently, the behaviour of birds was described in very woolly and imprecise terms even in erudite handbooks. The reasons for this vagueness are not difficult to guess at because the quaint rituals of birds often take place in the seclusion of foliage or in circumstances when privacy is guaranteed; the majority of bird watchers rarely have either the time or inclination really to watch their quarry, and more often than not succeed in making them nervous and anxious. And yet it is precisely the elaborate behaviour of birds in action, whether preening or courting, that makes them so fascinating.

This book attempts to expose something of the private life and the world of these creatures in reasonably straightforward terms. For the non-specialist, the subject has its difficulties; the results of careful scientific analyses are often as elusive as the behaviour itself, being buried in learned periodicals. No pretence is made that the text covers the total spectrum of bird behaviour studies; a line must be drawn somewhere and so I have tended to give a more thorough treatment of subjects that will be appreciated by the general reader (such as the language of displays and their origin). Although the book can be read from cover to cover, most double-page spreads deal with a single topic; there is a story on each page.

No matter whether you watch birds in the park or on wide open shores, feed them in your garden or keep them in cages, it is hoped that this book will help you to view birds with a greater understanding and appreciation.

J.S.

Published by The Hamlyn Publishing Group Limited
London · New York · Sydney · Toronto
Hamlyn House, Feltham, Middlesex, England
In association with Sun Books Pty Melbourne

Copyright © 1969 by The Hamlyn Publishing Group Limited

Phototypeset by Oliver Burridge Filmsetting Limited, Crawley, Sussex
Colour separations by Schwitter Limited, Zurich
Printed in England by Sir Joseph Causton & Sons Limited

CONTENTS

A Mockingbird singing

INTRODUCTION

The Fascination of Birds

There is probably no place on Earth that has not been darkened by the shadow of a bird; this gives one some idea as to how widespread birds are! There are more than 8,600 different species belonging to 166 families, and some, at least, have come to occupy quite a place in our lives. We feed them on our bird tables, keep them as pets, stalk them with binoculars and guns, wear their feathers, and enjoy their flesh and eggs.

Animals that move are always more fascinating than inert ones and birds score well here because they are very restless creatures. Furthermore, most birds can fly and who has not envied their freedom of the skies?

Surprisingly, no individual bird appears in the 'top ten' of our favourite animals; the penguin comes nearest, but, even so, it is a non-starter when compared with the chimpanzee or panda. However, birds as a class of animals endear themselves to us because we both share an appreciation of visual stimuli. Whereas most mammals live in a world of smells and have faces dominated by enormous noses in order to perceive smells, we, in common with higher primates, depend to a large extent upon our eyes. Our ability to appreciate forms, colours, and movements stems from this faculty and at the same time predisposes us to appreciate birds because, like us, sight is their chief sense. Accordingly, they have evolved displays that are full of visual impact; of course they have a special meaning to birds but we nevertheless find them amusing and colourful. Birds, like ourselves, also have a sound language. We find some of their voices exceedingly pleasant and it has even been suggested that birds such as nightingales and mocking-birds with sweet, melodious songs have an inventive or creative musical ability.

Their rich repertoire of behaviour makes birds interesting subjects for studying and this book will attempt to show something of this variety, to pry into the minds of birds, and to document some of the fascinating facts recently discovered.

The Life of Birds

Even while it is still inside the egg, a bird might not be altogether oblivious of the outside world; indeed it might even learn to react to its mother's call. When hatched, most birds depend upon their parents for food, warmth, and protection; once free it must keep itself clean, find enough food and roosting places, and avoid trouble with enemies. Later on it will act as mate and probably as parent itself to carry out the complex task of reproducing. Life will never be easy – failure usually means death and survival might depend upon the bird behaving correctly at the right moment.

Like human beings, birds depend much on their eyes. The Snowy Owl (*right*) has large eyes that are well adapted for nocturnal hunting. The fox (*opposite*), like most mammals, uses its nose in tracking its prey.

5

What is Behaviour?

Living organisms are characterized by the fact that they reproduce, grow, and often react; this reactivity usually takes the form of movement of the whole organism, responding and adjusting itself both to the external world and to its changing 'internal world'. For example, a sparrow will constantly have to adjust its balance when it is perching on a flimsy, springy twig, and if hungry, it will fly down to the ground in search of food. In the first case, external events are causing the bird to maintain its balance by exercising different sets of muscles; in the second case, flight is induced by internal events (hunger).

Behaviour is really the way in which an animal acts and it can often be demonstrated that it has *survival value*; the way in which sparrows respond to a prowling cat has a very immediate bearing upon their very existence. Those that react quickly stand a better chance of escaping and leaving more progeny than the tardier individuals. This weeding-out process of *natural selection* means that behaviour which *enhances* the survival and success of the species is preferentially handed on to subsequent generations; this, of course, chiefly

Behaviour need not involve movement. An alarmed Bittern adopts a vertical, motionless posture and is very difficult to see against the reeds.

6

Chaffinches and a House Sparrow escaping from a cat.

holds true of behaviour that has a genetic or inherited basis. The behaviour patterns we can observe today, then, have had a long history of evolutionary development.

Although movement is an essential ingredient of most behaviour, some behaviour involves giving it up altogether. The Bittern freezes in a special posture when threatened with danger, and, being well camouflaged, it seems to vanish into its reedy surroundings. This example illustrates well the fact that behaviour operates in circumstances that vary from species to species (e.g. with the kind of habitat and predators). The Bittern's best chance is to become immobile, but with fairly conspicuous birds such as sparrows or starlings, these tactics would be suicidal.

The term 'behaviour' covers such a wide range of activities that it is useful to recognize two categories of behaviour. General locomotion (such as flying), feeding and preening are all examples of *maintenance activities*, so named because their effect is to keep the individual in good order while having relatively little influence on other birds. Exceptions can always be found to this generalization; for example, the sight of other birds feeding may be sufficient to encourage others to feed, although they may be quite satiated. This phenomenon is called *social facilitation*, and is discussed elsewhere more fully. The other class of behaviour is concerned with conveying information to, and influencing the moods and activities of, other birds of the same species. These kinds of behaviour can be called *displays*.

The evolution of language codes have enabled birds to be social. Of course, some are far more gregarious than others. For example, many birds of prey come together in pairs in order to breed but otherwise they lead fairly solitary lives. On the other hand, many seabirds such as Adelie Penguins form great bird 'cities' during the breeding season and, like our own urban societies, there is a degree of orderliness about them.

An Adelie Penguin rookery

Discipline is achieved with rules, regulations, and conventions that each bird must observe or else it will be attacked, ostracized, and fail to breed.

Displays are often conspicuous, attractive to our eyes, and carried out in a stereotyped manner; it is not altogether surprising then that these have been the favourite study material of ethologists (scientists concerned with analysing behaviour). To begin with, there is always the challenge of cracking the language codes of animals, and of assessing the moods of the signallers. To do this, the rituals are carefully described and analysed, and the preceding and subsequent behaviours are noted together with the reactions of individuals that perceive the displays. The situation in which displays occur is also important. For example, a ritual that is always directed to an opponent and is more often than not followed by attacking, might convey the message: I am likely to attack you if you don't clear off. A posture performed on the nest site by an unmated cock that is followed by a hen landing by his side might be an advertising display; it could, however, equally be carried out for the benefit of other cocks to warn them off the territory that he has claimed as his own!

Acting on Impulse

Another fruitful field for research has been the investigation of stimuli that cause birds to behave in a certain way. A Robin defending its territory is more sensitive to some features in its surroundings than to others; for example, it will attack any strange Robins. The key stimulus that spurs the bird into action seems to be the opponent's red breast because even a bunch of red feathers may be mercilessly attacked as though it was a real rival. Simple sign stimuli that elicit compulsive reactions in certain circumstances play a fairly significant role in the lives of most birds. For example, brooding Herring Gulls recognize their eggs by a combination of stimuli, such as colour, shape, size, and markings. When they have a choice, these birds

A Herring Gull incubating a 'super egg'. A normal egg is shown beside the 'super egg'

prefer to incubate speckled to plain ones, rounded to angular eggs, and large to small ones. It is possible to make an egg with all the most effective features exaggerated – a 'super egg' with super-normal stimuli that a Herring Gull will compulsively incubate in preference to its own normal eggs.

Responding on the one hand to such a narrow stimulus or, on the other hand, to an absurd super-normal stimulus might seem strange to us and yet comparable examples can be quoted from our own experiences. Babies a few months old will smile at a sheet of paper with two painted black circles resembling eyes, and women with super-normal breasts litter advertisements and comic strips (*below*). Manufacturers tell us how effective these sign stimuli are for selling their wares.

We're
forward
looking
creators......

of large Engineering Projects

11

Swallows do not learn to fly although experience may perfect their movements.

Innate and Learned Behaviour

To what extent are actions predetermined by inheritance on the one hand and learning on the other? A few years ago most people held the view that much bird behaviour is governed entirely by instinct and relatively little affected by previous experience. This view is altogether too simple because investigations have shown that, although some behaviour is rigid and develops automatically, birds do show considerable learning ability. Examples will make this point clear.

Flight develops automatically according to a set of inherited instructions; young pigeons prevented from exercising their wings fly, when released, almost as well as young birds reared under normal conditions. In this, as in many other cases, practice makes perfect. As we shall see later, some birds like meadowlarks and blackbirds develop their own kind of songs without having to learn from others. The movements of copulation are inborn, although their correct orientation with respect to the mate does not come automatically.

When behaviour is largely determined by inheritance, it is extraordinarily inflexible although it may be singularly inappropriate; this point is well illustrated by experiments on lovebirds. Two quite different methods of transporting nest material are found in these birds. The Peach-faced Lovebird tears off strips of bark or leaves and tucks them in between the rump feathers before carrying them back to the nest site. Fischer's Lovebird carries its nest material in the bill. Hybrids between these two species are endowed with two sets of incompatible instructions for carrying; in their confusion, they try to place the material between the rump feathers but pull it out again because at the same time they feel the need to carry the material in the bill and refuse to let go (*right*). It takes them three years or so to abandon the rump feather technique in favour of carrying the material in the bill. Thus, when maternal and paternal behaviours clash in the hybrid young, learning is very slow to come to the rescue.

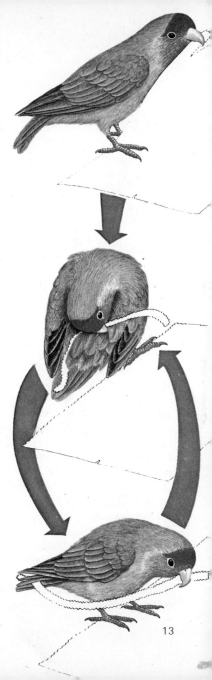

13

Newly-hatched chicks instinctively peck at small spots. Some of the spots may be edible and so by trial and error the young bird quickly learns what is food and what is to be avoided. The pecking action is innate but its use becomes more selective through learning until the bird can easily differentiate.

Throughout their lives, birds pick up information about their environment and companions, and this store of knowledge will help them to modify some of their behaviour according to their previous experience. Details of their home ground, such as where the best roosts are, what is good to eat and where to find it, which birds are to be avoided and which ones are to be bullied with impunity – this sort of information helps them to survive. In addition, special clues are automatically stored at certain critical periods during life, and these will be discussed later on.

In experimental situations, some birds do have the ability to solve problems. Goldfinches have for centuries been kept in special cages in which they could only get food and water by pulling up containers at the ends of pieces of string. Tits learn this trick very quickly although these birds are naturally very exploratory – they use the bill and feet in a co-ordinated way while feeding and so they are more likely to stumble on the correct solution. Chaffinches find it difficult to learn these techniques because they are quite alien to their way of life.

String manipulating by a Great Tit. The bird can only obtain the food by the deft use of the bill to draw up the string which is then successively held down by the foot.

KEEPING CLEAN

Cleanliness has survival value for all animals. Because of the very delicate nature of feathers, the toilet behaviour of birds is particularly complex to ensure that every feather is given the individual attention it requires; even a sparrow-sized bird has between 2,000 and 3,000 feathers. The feathers overlap in such a way that a layer of air is trapped next to the skin, helping to retain the body heat; in aquatic species it also acts as a built-in buoyancy tank. The pinions and tail feathers are specialized to produce the lift without which flight would be impossible. Birds also use their plumage in various ways to communicate; indeed, breeding is almost a matter of exchanging social signals and so any malfunctioning of this visual language system would unfavourably affect reproduction.

In the rough and tumble of everyday life, the plumage is likely to become deranged and contaminated with nectar, pollen and mud, and must be cleaned. In addition, and just as important, the surface of a bird provides a warm home for innumerable species of insects, some of which would decrease the hosts ability to survive if left unchecked.

Fleas and lice are perhaps the most important parasites that live in the plumage. Adult fleas suck blood from their hosts and although some lice feed on blood too, most of the 25,500 species chew up the feathers, particularly the fluffy barbs at their bases. Both kinds of insects are exceedingly sophisticated surface parasites; fleas have a resilient armour (exoskeleton)

A hen flea (magnified ten times)

A Common Louse (magnified five times)

and are laterally flattened so that they can slip easily between the shafts of the feathers. Lice cling to the feathers and are dorso-ventrally flattened to protect them from their host. A Robin with its upper mandible missing – and therefore unable to preen properly – had 127 specimens of the louse *Ricinus rubeculae*, whereas a normal count on a healthy Robin would have been about fifteen. This demonstrates the effectiveness of preening in keeping parasites at bay.

A Song Thrush with its plumage dirty and dishevelled through lack of preening.

Bathing

Bathing is a regular feature of a bird's life. Indeed, should the opportunity to bathe be withheld in captive individuals, they show extreme signs of deprivation, and lose no time in jumping into a shallow water trough as soon as they are given the chance to do so. In highly gregarious species, the mere sight of a bathing bird is sufficient to arouse a bathing mood in others. In aviaries, the contagious effect of this behaviour is only too easy to demonstrate, and if the bathing pool is too small to accommodate all of the birds at once, then they may queue up or jostle each other for a place in the water. Those that cannot gain access immediately may even go through the actions of alternately wetting and drying their plumage while queuing up! This 'infectiousness' of behaviour has the advantage of keeping social birds together, as under normal circumstances their chances of survival are better in flocks.

The immediate purpose of bathing is to wet the feathers; this action will tend to dislodge surface dirt from the plumage and therefore aid the process of getting clean. Nevertheless, it is possible that by soaking the feathers, the subsequent activities of preening and oiling may be made more efficient.

Not all birds bathe – those groups like the Galliformes (an order that includes the currasows, grouse, pheasants, turkeys, and incubator birds) do not but are inveterate dust bathers.

Aquatic species such as grebes, divers, gulls and ducks bathe while floating; they repeatedly duck beneath the surface, sometimes turning a full somersault, and then beat the wings against the flanks so that the upper parts are splashed. This action is followed usually by rubbing the head along the flanks and wings.

Some land birds tend not to use standing water for soaking their plumage but are rain or dew bathers. Birds such as parrots and hornbills have characteristic postures that expose

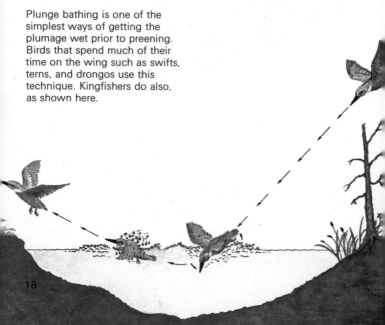

Plunge bathing is one of the simplest ways of getting the plumage wet prior to preening. Birds that spend much of their time on the wing such as swifts, terns, and drongos use this technique. Kingfishers do also, as shown here.

the maximum surface to the rain – the wings are spread out, the tail fanned, and feathers erected especially in the region of the oil or preen gland. Larks bathe only in the rain.

The majority of terresrial species tend to bathe while standing in shallow water with their feathers ruffled to allow water to penetrate the plumage. The precise movements vary from one family to another. Babblers, for example, repeatedly jump in and out, staying in the water only briefly.

Post-Bathing Preening

Preening consists of perhaps twenty different movements, some simple and others complicated, that are often carried out at great speed and in a more or less orderly sequence. Body shaking and wing whirring occur immediately after bathing to rid the plumage of excess water. The bill is then wiped before extracting oil from the preen gland, which is then smeared over the feathers. Full powers of flight must be restored as soon as possible and so the pinions are preened before the remainder of the plumage, which is dealt with in a more leisurely fashion. For finches, the whole session might take twenty

Most perching birds use two sets of highly co-ordinated movements while standing in the water. First the bird tips its body forwards, and with the head and breast immersed the bill is thrashed around and the wings vigorously flicked up and down.

Alternating with this movement, the body sinks back on the tail, and the wings are 'scissored' over the back so that the water is showered over the plumage. This is a Goldfinch.

minutes, ending with a vigorous body shake to settle the feathers correctly.

On these pages are illustrated a few of the *motor patterns* or actions that occur in the post-bathing preening sequence of a Red Avadavat, a small Indian Grassfinch (Estrildidae) which is often kept by aviculturists.

1. Beak wiping serves to clean the bill before preening.
2. Oil is taken from the preen gland. This will be scratched over the bill and then smeared over the plumage.
3. Five different actions are used to preen the pinions. Here the primaries are drawn through the bill.
4. Secondary and tertiary wing feathers are preened by twisting the head backwards between the axilla.
5. The scapulars and rump feathers can be reached because the neck is very flexible.
6. The tail has to be twisted round to clean the vent.
7. Preening the under-wing coverts.
8. The breast feathers may be left until ten minutes after the beginning of the sequence.

Oiling

Unlike mammals, birds have very thin skin with no sweat glands, whose secretion would otherwise foul the plumage. To make up for the comparatively non-glandular skin, most species of birds have a prominent gland situated just above the root of the tail, and usually covered by overlapping feathers; this is called the oil or preen gland and the waxy and fatty secretion from it is applied to the plumage during preening. Ostriches, emus, cassowaries, frogmouths, and bustards do not have one, and it is hardly functional in the pigeons, parrots, and woodpeckers. The fact that it is particularly well developed in aquatic species, and those whose plumage is regularly wetted, such as Ospreys, suggests that the secretion is used for making the feathers water-repellent; indeed, should the oil gland be experimentally removed in ducks, the condition of the plumage usually deteriorates.

It has been shown recently that when the oil gland secretion is irradiated by sunlight, Vitamin D is produced. If this happens on the bird's plumage then this substance could either be taken up by the bill

during normal preening bouts, or absorbed directly through the skin; indeed, there is evidence that birds develop rickets if their oil glands are removed.

Apart from water-proofing the feathers and providing a source of Vitamin D, the oily secretion may be used as a defensive odour in hen Hoopoes when they are incubating. The Musk Duck's name derives from the smell emanating from its oil gland. Also yellow, orange and red carotenoid pigments from the food are often secreted by the gland and this colouring is then applied to the plumage. The rosy bloom that occurs on the breasts of some gulls and terns during the breeding season originates in this way, and the Great Hornbill from Asia applies its yellow oil gland secretion to certain areas of the wings. Birds were using cosmetics long before man!

Scratching

The head plumage is of course inaccessible to the bird's own bill and so this area of the plumage is serviced by scratching. Also, many perching birds transfer oil from their bill to the head plumage by scratching. Species that feed on fish are likely to get their feathers soiled with slime; to combat this, herons have areas of down that produce a fine powder. The powder soaks up the slime, making it easy for the birds to remove by scratching with their special comb-like claws.

Two methods of scratching have been described depending

A male Mallard using its oil gland

Direct scratching — a
male Hyacinth Macaw

Indirect scratching — a
male Bullfinch

upon whether the scratching leg is brought straight up to the head (the direct method) or whether it is brought over the shoulder blade of the wing, which is especially lowered (the indirect method).

The majority of bird groups scratch directly, whereas the indirect scratchers include most of the perching birds, turacos, swifts, hummingbirds, Hoopoes, oystercatchers, avocets, and stilts. Both indirect and direct methods have been described in the American wood warblers and parrots.

Bronze-winged Manakins are birds that preen each other. Social preening occurs in birds that rest flank-to-flank in groups, such as grassfinches, parrots and babblers.

Male and female Oriental White-eyes allopreening. The body feathers are fluffed up when inviting preening.

24

Cuban Finches allopreening. The cock (*left*) invites preening by holding its head up and raising its feathers.

Preening Each Other — Social Preening or Allopreening

Social preening or allopreening is scattered throughout thirty-eight families of birds from penguins to crows; this behaviour takes the form of one individual preening the plumage, usually on the head, of another and it sometimes does this in response to special preening invitation postures. These usually involve ruffling of the head plumage (more general body ruffling in the case of the white-eyes) and perhaps facing away so that the neck is presented to the neighbour; when being preened, an inviting bird may adjust the position of its head so that all of the feathers are worked over by the preener.

Although the social implications of this behaviour will be dealt with later on, there is reason to suppose that the head does provide a safe refuge for parasites where they will be out of reach of their host's searching bill, and of course out of sight. The body lice of birds like swans and Golden Orioles are white and yellow respectively, but those species of parasites that pass sedentary lives in the head plumage are not camouflaged. The fact that they can flaunt themselves shows that they are in little danger of being discovered, at least in species that have not evolved allopreening. Preening of one individual by another may then be important in preventing parasites from accumulating in the head region.

Allopreening occurs in birds that sit flank to flank and between pairs of birds that have to share small nest sites, such as Noddies, Fairy Terns, and Razorbills.

Anting

Anting is the most curious of all feather maintenance activities and is confined to perching birds (Passeriformes). Like bathing and preening, the behaviour is highly stereotyped and involves either adopting postures that allow ants to stream over the body or applying ants directly to the plumage. There is a preference for those ants that squirt formic acid and it is possible that its insecticidal properties might kill off parasites, particularly on the wing tips, which are all but inaccessible to the bill. Substitutes such as cigarette ends, lemon juice or even beer may be used by captive birds.

House Sparrows take elaborate dust baths. They scratch with the feet, as when constructing nests, and make little bowls in the fine soil.

The Jay is a passive anter and adopts this posture in response to wood ants. So much formic acid may be squirted by the ants that the plumage glistens.

A Starling actively applying a mass of ants to the wing feathers. There are species that manipulate single ants until they are of no further use.

Dust Bathing

As birds spend such a great deal of time cleaning themselves, dust bathing might at first sight appear nonsensical! However, live lice have been taken from dust baths regularly used by chickens, and so this behaviour might help to rid the body of unwanted parasites. Partridges and pheasants deliberately kick soil onto the back and sift it through the plumage.

Insects form an important part of the diet of many birds, and some have developed defence mechanisms. The Jay is startled by the eye-spots on the wings of a moth.

FEEDING

Finding food of the right kind and in sufficient quantities is literally a matter of life and death to birds, particularly during the winter when food may be in short supply and the hours of daylight are minimal. For example, in scientific observations of the feeding habits of several birds, it was discovered that in winter tits must find an average-sized morsel of food every $2\frac{1}{2}$ seconds for 91 per cent of the day, and Wood Pigeons must search for food for 95 per cent of the daylight hours, otherwise these birds will starve. Although the behaviour patterns for feeding may be inborn, these become more efficient with practice and experience. Furthermore, birds no doubt learn where food is likely to be found and they may even come to learn what to look for as a result of experience.

Swallows drink on the wing by scooping up water.

Drinking

Apart from a few desert species, all birds need to replace the water that they may lose in their breath and excreta. Antarctic penguins will make this loss good by eating snow but the majority of birds living in warmer climates will drink. Species that fly most of the time take their water while on the wing, but most birds drink with their feet firmly planted on the ground. Water is scooped up in the lower mandible and then made to run into the throat by tilting the head backwards. Pigeons and some desert inhabiting grassfinches suck water into the crops without tipping and this may be an adaptation for making use of small quantities of fluid.

There is a certain amount of evidence that birds have no inborn recognition of water but must learn to recognize water and to drink it.

Greenfinches drink by taking water in their bills and tipping it back down their throats.

Wood Pigeons drink by sucking up water.

The Variety of Beaks

The beak may simply act as a pair of forceps, grasping the food before it is swallowed, as in flycatchers, but sometimes it is used as a probe, as in waders. More usually it functions as an instrument for processing the food before it is ingested, as in husking seeds or ripping flesh. Beaks may even provide sites for special colour signals denoting sex or species.

Insectivores such as the Nightjar have weak bills and enormous gapes surrounded by bristle-like feathers that help to contain their prey.

Geese have bills with serrated edges for cropping grass. This is a White-fronted Goose.

Graminivorous birds use their bills to husk or crack seeds, and the larger the seed they feed on, the heavier the bill and stronger the jaw muscles. Hawfinches (*left*) can even crack cherry stones. Crossbills (*right*) have special beaks to extract the seeds from pine cones.

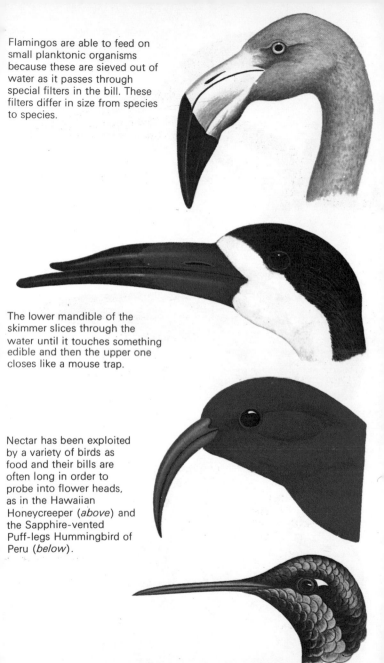

Flamingos are able to feed on small planktonic organisms because these are sieved out of water as it passes through special filters in the bill. These filters differ in size from species to species.

The lower mandible of the skimmer slices through the water until it touches something edible and then the upper one closes like a mouse trap.

Nectar has been exploited by a variety of birds as food and their bills are often long in order to probe into flower heads, as in the Hawaiian Honeycreeper (*above*) and the Sapphire-vented Puff-legs Hummingbird of Peru (*below*).

The Greater Spotted Woodpecker has a chisel-like beak for uncovering wood-boring insects.

The Huia, an extinct New Zealand bird, was the only species in which the sexes had different kinds of bills and this enabled them to exploit separate sources of food. The male had a chisel-like beak, like the woodpecker, whereas the female probed into crevices with her longer and more delicate pair of tweezers (*opposite*).

The Common Kingfisher has a long pointed bill for spearing and holding fish.

Most birds of prey have strongly-hooked bills for shredding their prey. The rare Everglade Kite specializes in feeding on water snails and uses the long tip to impale and then extricate the fleshy body from the shell.

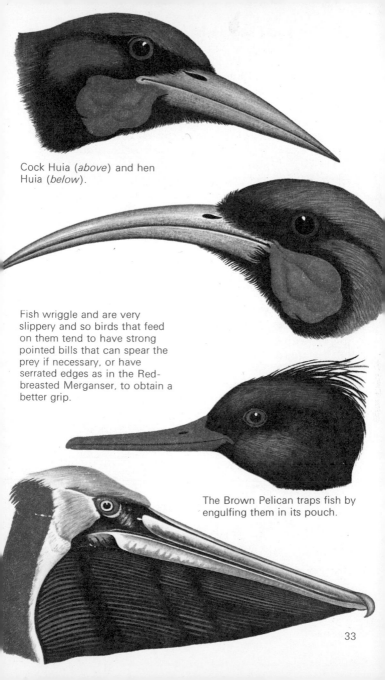

Cock Huia (*above*) and hen Huia (*below*).

Fish wriggle and are very slippery and so birds that feed on them tend to have strong pointed bills that can spear the prey if necessary, or have serrated edges as in the Red-breasted Merganser, to obtain a better grip.

The Brown Pelican traps fish by engulfing them in its pouch.

33

Bill Size and Feeding Habits

The way in which a bird's anatomical equipment imposes certain limitations on the choice of food and the role experience plays in reinforcing it have been investigated for seed-eating finches. Before the seed is swallowed, it is taken in the bill and either completely husked or else the outer protective layers are split, the nourishing kernel being scooped out with the tongue. Finches, however, differ from one species to another in the length, depth, and width of the mandibles. Seeds also come in all sizes from millet (200 seeds per gram), through hemp (40 seeds per gram) to sunflower (7 seeds per gram).

If given a choice of these seeds, Greenfinches take sunflower and hemp in almost equal proportions, whereas Linnets prefer lighter rape, canary, and millet seed in that order. The interesting point is that the more delicate-billed species will, within reason, tackle the stouter seeds that provide large juicy kernels once they have been dehusked. This process would take so long for small-billed kinds that it is more economical for them to select smaller varieties; a Chaffinch takes on average forty-five seconds to deal with a sunflower seed whereas the Greenfinch deals with it in ten seconds. So species with small bills work at greater efficiency in husking smaller seeds. The advantage of a stout bill is that sufficient food can be obtained by opening relatively few but large seeds; a Waxbill living on millet would have to eat 336 seeds whereas a Greenfinch living on sunflower need only take fifteen seeds to obtain one gram of food.

A Chaffinch does not have a strong bill and prefers to eat medium-sized seeds such as hemp.

A Linnet has a lighter and more delicate bill and eats large quantities of small seeds such as millet.

The young of large-billed finches might select small seeds at first but they gradually learn to husk the larger ones more efficiently than their smaller-billed relatives because they would have less difficulty in holding them. The need for greatest efficiency in feeding (in terms of weight of food taken per unit of time) leads to differences in diet of seed-eating finches and this is governed by a combination of learning and bill size.

A Greenfinch has a stout bill that can easily open large seeds such as sunflower, of which it is very fond.

The Use of Feet in Feeding

The very act of walking through vegetation may cause insects to stir and therefore be discovered. Certain herons, gulls, and waders may rhythmically stamp or mark time on the ground or in shallow water and this behaviour undoubtedly helps them to find food.

The feet are often used in conjunction with the bill in preparing food. In the carnivorous raptors (birds of prey), the

35

The Blue Tit often holds pieces of food in its feet.

prey is securely held in the talons while the bill is used as a hook for preparing the food. Although the majority of birds simply take hold of the food with the bill, some grasp it first with the foot before transferring it to the bill. The ability to use the feet in this way lends a certain charm to birds like parrots, owls, and titmice. The latter undoubtedly use the feet to pull branches within range of the bill; this behaviour is put

The Purple Gallinule also uses its feet in feeding.

to good use on the bird table where peanuts suspended on a piece of string can be efficiently pulled up by deft use of bill and feet (see page 15).

The Tool Users

Only four species of birds use tools in the strict sense of the word. The best known example is that of the Short-billed Woodpecker Finch from the Galapagos Islands, which uses a cactus spine for routing insects out of deep crevices. As soon as one is flushed, the cactus spine is temporarily laid aside while the insect is snapped up. By using a tool, this insectivorous bird can take advantage of an otherwise inaccessible source of food. The ability to feed in this way must be inborn although experience may be necessary to use the cactus spine effectively.

Recently it has been found that Egyptian Vultures use large stones in order to crack open Ostrich eggs. The bird may select a heavy stone, pick it up in the bill and walk over to the egg, then rising up on its toes, drop the stone on to the shell. Immature birds may perform this behaviour in response to seeing an Ostrich egg, letting the stone fall to the ground away from the potential food source. This points to the behaviour being inborn, requiring experience to orientate it correctly.

The Woodpecker Finch of the Galapagos Islands uses a cactus spine for obtaining insects from deep crevices.

The Egyptian Vulture drops stones on ostrich eggs to crack them open.

A European Song Thrush breaks open a snail on a stone; a Blackbird looks on.

Three species of bower birds use plant pigments to adorn their bowers and two apply the 'paint' with brushes. A wad of fibrous bark is used by the Satin Bower Bird whereas the male Regent Bower Bird uses a pad of greenish leaves.

Of course there are species that use natural features of their environment in order to process their food. Although not examples of tool users, they are nevertheless worth considering. European Song Thrushes shatter snails' shells against special stones (anvils) to expose the succulent bodies. Gulls will drop hard-shelled crabs or molluscs from the air to crack them open. The Old World Lammergeyer, a graceful vulturine bird, has special dropping areas for breaking open bones to expose the nourishing marrow; tortoises may be treated in the same way. Nuthatches and Great Spotted Woodpeckers use crevices as vices where thick-shelled nuts can be hammered open by their powerful bills.

Storing Food

The amount of food available varies from season to season, and some species hoard food when there is plenty of it about. Shrikes impale beetles, lizards, and nestling birds on thorns to form larders although, as they are generally migrants, the stores cannot be for their long-term needs. The food that is concealed is probably of great survival value if it is found during the winter when food is scarce. Many European species of tits hide food away in crevices and clumps of lichen or moss thereby dispersing it throughout the living space. Nuthatches do the same and the North American Acorn Woodpecker may store up to 50,000 acorns in a single tree, boring out special holes to accommodate them (see page 41). The traditional storage tree is the focus of attention of these woodpeckers, which live in small communities.

The members of the crow family are also inveterate hoarders. Thirty or forty Jays in an English oak wood were estimated to have buried about 200,000 acorns during one October. The spread of oak trees, particularly up-hill, is dependent upon the industry of Jays and squirrels, and their occasional forgetfulness! In North America, the Grey Jay stores pellets by sticking them to branches or inside crevices with saliva from its enlarged glands; deep snow cover almost necessitates storage above ground in this species. Others build up caches beneath the surface, patting down the earth and often placing a leaf over the top for camouflage. Although tits and nuthatches probably have no specific memory for locating their stores, they work the area so thoroughly that up to 90 per cent of the hidden food is found. Jays and Nutcrackers can easily locate their larders of acorns and hazel nuts, often beneath several inches of snow, with amazing accuracy.

In Sweden, the Thick-billed Nutcracker, a member of the crow family, harvests hazel nuts. During the autumn when

The Grey Jay of North America sticks seeds and pieces of meat to trees with its saliva.

he nuts ripen, these birds work hard collecting this food, illing their throat pouches and flying off into their spruce orest territories where they deposit the nuts in little caches below the surface of the soil. When winter comes, the birds are able to live off their food stores, which last into the spring as they even manage to nourish their young on hazel nuts stored the previous autumn. Nutcrackers may be able to ocate their caches by remembering their locations with reference to conspicuous landmarks, such as trees or buildings, that remain unchanged by blankets of snow.

The Acorn Woodpecker of the United States stores acorns by inserting them into holes bored into trees.

The Thick-billed Nutcracker easily locates caches of nuts buried beneath the snow. In an observation of 351 nut stores covered with up to 20 inches of snow, 86 per cent. were recovered.

Feeding Associations

Food is where you find it, and it is often situated in rather unusual places. Mutually beneficial partnerships have been struck up between different kinds of animals; for example, Egyptian and Spur-winged Plovers have been credited with removing debris and leeches from the opened mouths of basking crocodiles; although this is probably a myth, there are equally fascinating feeding relationships. African tick birds or ox-peckers live for most of the time on the bodies of large

Greater Honey Guides draw the attention of honey badgers to bee or wasp nests, and wait their turn for the honeycomb.

mammals such as rhinos and giraffes, feasting on blood-gorged ticks and flies. They may even peck at scar tissue forming around open lesions. As their name suggests, the Scaly-throated and Greater Honey Guides lead honey badgers or humans to wasp or bee nests and wait for scraps of honey-comb left by the badger. Originally, these birds may have only scavenged opened nests but now the birds take the lead.

Many cases have been recorded of birds using larger

creatures as 'beaters'. Carmine Bee-eaters often perch on the backs of Kori Bustards or Abdim's Storks and await insects to be flushed from under the feet of their hosts. Foraging tribes of monkeys may also be followed by birds such as hornbills, drongos, and trogons. Various egrets associate with cattle and it seems likely that they use them as beaters. One study confirmed that Cattle Egrets following these large browsing mammals fed at $1\frac{1}{2}$ times the rate of, and took only two-thirds the number of steps of, birds that were hunting insects and reptiles by themselves, thereby taking more food and expending less energy. Nowadays, ploughs may be followed by egrets and gulls, and the opportunism shown by these birds illustrates how man is becoming involved in these feeding

Robins often watch men digging and snatch up exposed worms.

relationships. We also deliberately supply birds with food; during the lean months of winter, scraps and commercial preparations are placed upon bird tables and this feeding must help many birds to survive, such as Great Tits.

Birds have also taken full advantage of our own wastes. The increase in gull populations may be due to the glut of edible

Cattle Egrets with African cattle.

A Blue Tit raiding a milk bottle.

refuse churned out on to exposed rubbish tips. However, the species that has profited most is an unlikely one – the Fulmar, a relation of the albatrosses. Since the middle of the eighteenth century, its numbers have increased bountifully because of the supply of easy-to-catch fish and whale offal thrown overboard into the North Atlantic by fishing and whaling boats.

A flock of Waxwings may rob a bush completely of its berries before moving on to another.

SOCIAL ORDERS

Living Together or Apart Outside the Breeding Season

The old proverb 'Birds of a feather flock together' is true for many species. Others, however, like shrikes or most birds of prey, are essentially solitary during most times of the year and so the question naturally arises as to why 'togetherness' has survival value to some and not to other kinds of birds?

Take the problem of avoiding enemies. An individual in a flock benefits from the vigilance of a large number of birds and it is less likely to be taken unawares. But the camouflaged plumage of birds that feed on the ground conceals the bird best if it is solitary.

Every bird must also find enough food to live. It seems likely that the different degrees to which birds live together are related to the way in which various kinds of food are distributed throughout the environment. In cases where the food is *locally* abundant, as in the case of fruits, then species depending on this source tend to flock *if* their feeding methods do not interfere with each other. To those species that depend upon stealth, speed and skill to procure their food, then flocking might be a hindrance, because each bird may well keep disturbing the prey of another member of the flock. Occasionally, even for insectivores, togetherness may be advantageous as each individual may act as a beater for one of the others.

Solitary birds such as Skylarks are well camouflaged for protection.

Contact and Distance Species

Parrots and grassfinches tend to sit in huddles with their flanks pressed against each other. However, the more usual situation found in social birds is for the individuals to sit just out of range of the neighbour's bill; further closure is prevented by threat or avoidance behaviour. For any given species this minimum closure distance is more or less constant and results in the regular spacing of resting individuals (e.g. starlings sitting on a telegraph wire). For small species, the closure distance is about six inches and can be easily measured. This is called the *individual distance*.

Of course, in those species that sit in huddles (*contact* as opposed to *distance* species), the individual distance is zero. Most contact species are found in the tropics (e.g. grassfinches, parrots, babblers) and in these birds, heat conservation would not seem to be an important consideration. Keeping warm might have great survival value in colder climates particularly at night; wrens, tree creepers, Long-

tailed Tits and swifts sometimes cluster at night, presumably for warmth. Adult King Penguins from the sub-Antarctic space themselves out but their close relatives, the Emperor Penguins from the high Antarctic, cut down their heat loss by one-sixth by huddling when the weather is bad. In this way, they save valuable reserves of fat.

Peck Order

Birds live in a competitive world, and for them, life may be rather like the proverbial rat race with success coming to those that can bully their neighbours. Thus, the stronger and more aggressive birds always gain the best perches, choice food and mates in competitive situations with their less assertive flock companions. Consequently, each individual

Starlings space themselves out about four inches apart when resting, close enough to jab at each other.

has its own place in the flock and accepts that it must give way to some individuals that rule the roost or else be attacked; however, other less aggressive birds will promptly make way for it. This is the way peck orders or dominance hierarchies are set up, and after the initial jostling for power, the relationships become fairly stable. Even in aviaries this is seen in some birds.

A peck order of the kind seen in chickens or pigeons need not be a straight-line order of dominance (as shown diagrammatically, *left*); triangular relationships might exist so that A has precedence over B, and B over C, but A might have to give way to C. Furthermore, a bird's dominance over another might alter from day to day and can certainly be changed experimentally by the injection of hormones such as testosterone (male sex hormone) that influence aggressiveness; a weak, submissive bird that is bossed around by its neighbours may be turned overnight into a despot.

Although a peck order is established by aggression, hostile behaviour is subsequently reduced; order prevents squabbling, because each individual recognizes his position, accepts it, and does not constantly have to fight to maintain it. When food is very scarce, the low-status birds stand a better chance of surviving if they leave the flock and certain starvation to find food elsewhere; if they do, they will not have to compete with high-status individuals.

When a number of species commonly join forces and hunt together in mixed flocks, there may be a peck order between them. This can be easily observed on bird tables where competition for the available food may be high. For example, in Britain, Starlings can always get to the food first because they are large and domineering and can consequently see off House Sparrows which in turn have precedence over Great Tits, the most dominant of all the tit family. Blue Tits come next in the peck order and Coal Tits are the most timid of all (*below*).

Behaving Alike

A flock not only shows some kind of structure, but the behaviour of its constituent members is roughly synchronized so that the group's activity shows well-developed rhythms. At any one moment, most of them will be engaged in the same kind of activity, e.g. bathing, feeding, resting or preening. If one bird starts to bathe, then the other members of the flock might quickly take up this activity. This tendency to follow suit or copy each other's general behaviour is called *social facilitation*. Even humans exhibit this phenomenon. Yawning and laughter are easily triggered off in this way, and it takes a

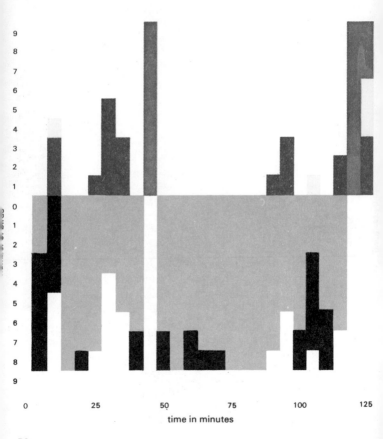

time in minutes

very resolute frame of mind to suppress these actions when faced with a yawning or laughing person.

Red-billed Weaverbirds are intensely gregarious and, therefore, have a well-developed tendency not only to be near others of the same kind but also to conform in behaviour. This is shown in the group activity chart on the opposite page, which shows diagrammatically the activities of nine Red-billed Weavers for just over two hours. The chart on this page shows the activities of six Bishop Weavers for the same period. By comparing these charts, which were made by Dr John H. Crook, it can be seen that the Bishop Weavers are less social than the Red-billed Weavers; they do not show such a marked degree of social facilitation and the individuals behave more independently of each other than do more gregarious species.

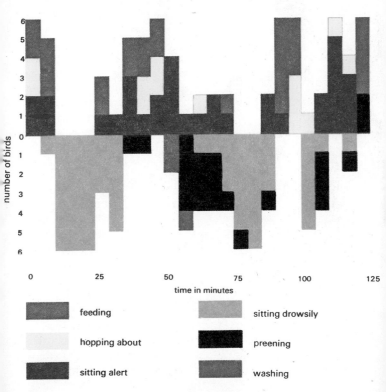

number of birds

time in minutes

feeding

sitting drowsily

hopping about

preening

sitting alert

washing

THE LANGUAGE OF BIRDS

Birds interact and respond to each other and when this happens there is usually an exchange of signals; these really constitute a *language*. Both visual and sound signals may be used for communicating although one or other might predominate depending on the species. A bird's system of social signals is quite different from our own; for example we depend a great deal upon facial expressions whereas birds convey subtle changes of mood by erecting part of the plumage or adopting certain body postures. Flash markings may be revealed by spreading the wings and so on. These behaviours make up a code of signals for communicating moods like anger or fear, or intentions, such as attacking or escaping or willingness to fly. Of course the precise details of the language will differ for every species but each signal will nevertheless be unambiguous to con-specific birds.

Cartoonists give birds muscular, expressive faces to make them communicate the 'human' way. In fact they have a much different 'language'.

Species Recognition

Every species has its own special markings and these might help to bring about spacing, in the case of solitary kinds, or aggregation in the case of gregarious species. The iridescent bands of colour on the wings of most dabbling ducks and the wing-tip and underwing patterns of gulls and birds of prey respectively are as much recognition marks as the national markings on aircraft wings.

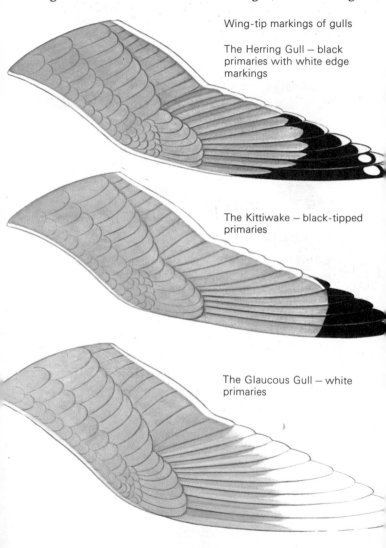

Wing-tip markings of gulls

The Herring Gull – black primaries with white edge markings

The Kittiwake – black-tipped primaries

The Glaucous Gull – white primaries

Conspicuous markings might tend to attract predators, but dull social signals that are useful in maintaining flock cohesion or spacing would hardly be effective because social signals, like those we use in our own society such as traffic lights, must be clear and unambiguous when given. A compromise has, therefore, been evolved by many social species: they have flash markings on the tail, rump, or wings that remain hidden unless exposed by lifting the wings or in flight. Many signals

Willet

Many birds have conspicuous markings that enable them to recognize others of the same species.

that aid in promoting cohesion or dispersion of birds are brought into display by wing movements.

In a few cases, conspicuous markings and behaviour are also useful in trying to attract the attention of predators when danger threatens a nest. Parent Oystercatchers, for example, will run ahead of a predator and pretend to be injured, thus luring the preying animal away from the real nest.

Whenever different kinds of birds fly and feed together, it would be an advantage for them to share, in some part at least, the same language or system of visual signals. A few cases have been described. Certain unrelated South American seed eaters that form mixed flocks have similar plumages and this no doubt has survival value.

American Avocet

Oystercatcher

In the wading birds shown here, these markings are mainly displayed in flight and hidden when the bird is on the ground or in the water.

Sound Signals

Many people will be familiar with the fact that birds use their voices to communicate information about their moods, such as whether they are angry, afraid or alarmed, etc., and status, for example whether they are unmated or mated. Each species has a repertoire of sound signals that differ more or less from those of all other kinds; they therefore serve the purpose of species identification. The number of different utterances varies from one species to another; Black-capped Chickadees are reported to have sixteen, the Song Sparrow and Chaffinch have twenty-one and the Great Tit may have as many as double this figure. Between one dozen and two dozen seems to meet the needs of the majority of birds.

The sounds that birds make can be divided into two categories: *call notes* and true *song*. The latter will be dealt with on page 104.

The simple call notes convey all kinds of information. *Contact notes* for example simply say 'Here am I, where are you?' and may be taken up by other members of the species

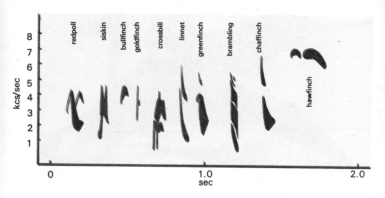

Some birds derive considerable benefit from living in mixed flocks outside the breeding season. In order to promote togetherness, activities such as feeding and flying have to become co-ordinated; this is helped by the evolution of flight calls that have an overall similarity while retaining the individual stamp of each species. Thus small finches, such as Siskins and Redpolls, which flock and feed together, are able to understand a part of each other's language.

within earshot and are, therefore, important in maintaining flock cohesion. The function of keeping the members of a flock together is also carried out by *flight calls*; these probably arouse others of the same species to fly and follow the caller and, therefore, help to generate a flying mood. Contact notes and flight calls certainly help nocturnal migrants to keep in touch with each other.

Sound Pictures

The calls and songs of birds are best displayed as sound pictures or *spectrograms*. As light often consists of a mixture of colours that can be separated out by a prism, so sound can be analysed into its component pitches at any instant by special equipment. This equipment records the sound in the form of a pattern; using this technique the differences between songs and calls can be readily appreciated at a glance. The higher the pitch, the higher the frequency on the scale; middle C is equivalent to 256 cycles per second and the upper limit to human hearing is seventeen to twenty kilocycles per second.

A wolf whistle looks like the lower spectrogram, and its form can be easily recognized. The upper one shows a flight call and an aggressive 'tseep' of a Red Avadavat (*right*).

59

Communicating Danger

Sometimes the calls of different species are astonishingly alike, particularly those that are intended to draw attention to predators. Furthermore, alarm calls given in response to a flying enemy are quite different from those elicited by the presence of a perched bird of prey, a fox or even man. In the latter cases, the predator is openly mobbed and the idea is to

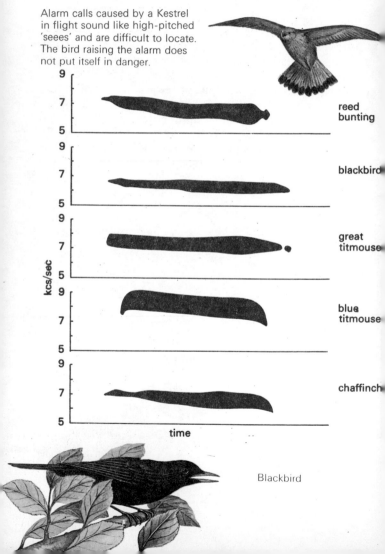

Alarm calls caused by a Kestrel in flight sound like high-pitched 'seees' and are difficult to locate. The bird raising the alarm does not put itself in danger.

reed bunting

blackbird

great titmouse

blue titmouse

chaffinch

kcs/sec

time

Blackbird

et all other birds in the district know where the source of danger is. Mobbing calls are therefore easy to locate. Aerial predators, however, are much more dangerous, particularly to the bird that first sounds the alarm because it draws attention to itself. These calls cause other birds to flee for cover but are 'designed' so as not to give away any directional clues to the hawk or falcon.

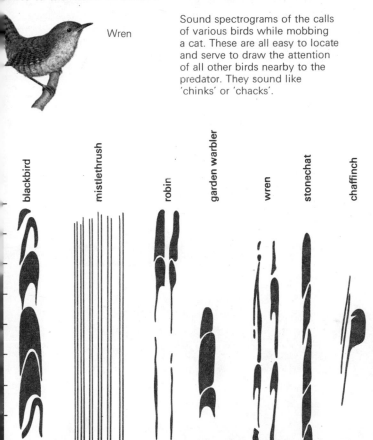

Wren

Sound spectrograms of the calls of various birds while mobbing a cat. These are all easy to locate and serve to draw the attention of all other birds nearby to the predator. They sound like 'chinks' or 'chacks'.

blackbird

mistlethrush

robin

garden warbler

wren

stonechat

chaffinch

time

Hostile Behaviour

Many of the every day interactions between birds involve aggression or avoidance behaviour. The bare necessities of life, such as food, perches and so on, must be competed for and success may be achieved by threatening a rival. Actual combat is a relatively rare event, however, and disputes are usually resolved by an exchange of ritualized threat displays that are designed to make the opponent frightened and withdraw.

The postures adopted by threatening or sparring birds reflect how angry or frightened they are; a confident and very

A cock Herring Gull threatening. The posture indicates that the bird *is likely to attack*: the bill is ready to strike downwards and the wings are held away from the flanks in readiness to beat the opponent.

A couple of Long-tailed Parson's Finches bill fencing

hostile individual will behave quite differently from a fearful defensive one, and so rivals can interpret each other's moods and adjust their behaviour accordingly. The significance of postures in the threat code can be followed in the encounter illustrated, between a cock Crimson Finch and hen Zebra Finch where the Crimson Finch is the aggressor.

When two birds are equally balanced, and neither backs down, then the preliminary exchange of threats may lead to fighting. In finches, the bills are used and attack behaviour tends to take the form of bill sparring or bill fencing.

The hen Zebra Finch has come rather too close to the rather belligerent cock Crimson Finch. The horizontal body posture of the Crimson Finch indicates a high degree of confidence and aggression.

The upright stance of the hen Zebra Finch shows that she is really rather defensive and frightened, although she is pecking back.

The encounter draws to its inevitable conclusion. The dominant and aggressive Crimson Finch succeeds in making the Zebra Finch fly away.

Aggressive displays are shown on the left — appeasement ones on the right.

European Jay

The bill-up posture shows a readiness to fly up and withdraw.

Sandwich Tern

During threat, the bird's weapon is pointed at the rival.

Black-headed Gull

Head flagging is an appeasement ceremony. The bill and intimidating mask are hidden from the opponent.

Appeasement Behaviour

Special social signals have been evolved that help to reduce aggression in others; these appeasement postures are just as important as threat behaviour, which is designed to increase fear. A bird that adopts a submissive or appeasement posture indicates its readiness to withdraw from a conflict.

Appeasement ceremonies are in a sense the opposite to those of threat; the bill is the bird's chief weapon and is pointed at a rival during hostilities, but appeasement postures often result in an exaggerated movement hiding the bill from the opponent's view, or pointing it upwards.

Certain displays given by losing individuals seem to be infectious and thereby reduce the chances of physical combat occurring. Communal displays like this have been reported for Jackdaws, gulls, Swallow Tanagers and Oystercatchers.

Some species exhibit displays that serve to frustrate aggressive tendencies in an opponent by arousing other drives that compete with them; social preening invitation postures seem to work in this way. These displays are found chiefly in contact species, which sit in huddles; as sitting pressed up to flock mates is a situation likely to cause irritability, the evolution of a powerful appeasement gesture that operates in these circumstances, particularly the ruffling of head feathers, helps to keep the peace by eliciting preening.

Communal displays help to prevent aggression, as in the piping of Oystercatchers (*above*). Social preening has a similar effect in some birds, such as Australian Gannets (*right*).

THE BUSINESS OF BREEDING

Animal populations must produce enough surviving young in order to compensate for the annual loss of adults, or else the species will eventually die out.

During the annual breeding seasons, male birds must compete for territories, a mate must be attracted and courted, nests built, eggs incubated and enough food collected for the ravenous young (which must be protected from equally ravenous predators). All of these activities require co-operation between the pair, at least for the majority of birds.

At this time of year, birds' behavioural repertoire goes through an enormous expansion. Individuals of opposite sex that would have squabbled over food and otherwise ignored each other, will come together and co-operate. This change is induced by hormonal developments that bring about a whole language of courtship; the exchange of these special signals makes possible the establishment of territories, mate attraction and the gradual loss of aggression and fear between the

Yellow-eyed Penguins form long-lasting pair bonds. One pair bred for thirteen consecutive years.

male

female

The male Ring-necked Pheasant (*above*) and hen
Painted Snipe (*below*) are both very brightly coloured and take no
part in rearing the young as they would draw attention to the nest.
The Painted Snipe may mate with several cocks.

partners so that chicks may be reared safely and successfully.

The majority of birds split up into pairs to breed. However, some form themselves into promiscuous clans and, in species where the male plays no part in rearing the young, polygamy often reigns. With some birds, the reverse occurs and the cocks take over the rearing of the young.

male   female

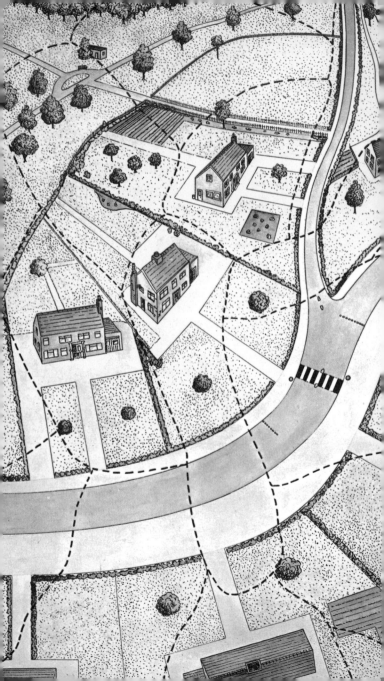

Territory

The breeding activities of most birds centre on defended areas or territories, which vary in size from one species to another. In most perching birds it may be an acre or two in extent, but the Golden Eagle may command an area of thirty square miles. As a result, the population of each species becomes dispersed throughout the suitable habitats during the breeding season. This spacing-out is a direct consequence of the antipathy of the cocks who, having established themselves, vigorously exclude other cocks from their territories.

The friendly Robin of English gardens is by no means benevolent to other Robins. During the spring, cocks disport their flame-red breasts in threat to each other (*below*); indeed, so powerful is the stimulus of 'redness' that even cock Chaffinches with pinky-red breasts or a bunch of red feathers will be savagely assaulted without hesitation if placed in a Robin's half-acre territory.

Opposite, the territorial limits of a number of pairs of Robins. Both cocks and hens protect the boundaries and drive off intruders. After the breeding season, the hens wander off and set up their own winter home ranges.

What is the survival value of territorial behaviour? It certainly provides a refuge for courting and mating and, in species like the Robin, it might also safeguard an exclusive food supply. But sometimes territories appear to be larger than is necessary for just providing food. In Robins, the defended areas become smaller when the population is larger than normal but the breeding success if not significantly affected.

Oystercatchers may have two territories; one inland where the nest is situated and another on the shore where all of the feeding takes place. When the young are being fed (they are one of the few waders that care for their chicks in this way), the parents commute between the territories. Sociable sea birds use their small nesting areas as a focus for courtship and chick rearing, but they forage far and wide in search of food and do not have exclusive feeding territories. Many finches and aerial insect hunters, such as swifts and martins, similarly have no proper feeding territories, and even during the breeding season many birds feed socially away from their exclusive nest sites – gulls are an example.

Territorial aggression does bring about a spacing of individual pairs but, at the same time, there are attractive forces operating so that territories tend to adjoin each other and become arranged into neighbourhoods. The degree to which the breeding areas are concentrated depends upon the survival needs of the species (for example, whether or not the territory has to encompass a food supply). Black-headed Gulls' nests tend to be spaced out one yard apart and research has shown that this distance is advantageous. Should the nests be closer then foraging Carrion Crows can more easily find the eggs; if the nests are further apart then it is possible that the concerted mobbing of predators would be less effective, because the members of each pair only mob when the predator is more or less infringing their own territory. The degree of spacing is, therefore, a compromise between conflicting demands.

Highly gregarious species, such as gulls, terns or herons, often nest on protected sites where the dangers of nesting in conspicuous bird cities that attract enemies are reduced. These sites include islands, isolated peninsulas, cliffs and trees.

A colony of Black-headed Gulls

Attracting a Mate

Every territory owner must establish his authority by repelling other males and at the same time lure a mate. The fact that there is unrelenting competition for suitable areas has been shown experimentally by removing resident cocks; their place is soon taken either by a neighbouring one or by an altogether new bird.

A territory owner possesses a certain psychological advantage over rivals within its own plot and its presence alone might tend to avert trespassers. However, self assertion and advertisement can be carried out in a number of ways. Song, as will be described later, is a two-edged weapon and species that live in dense vegetation or under the cover of darkness use vocalizations to proclaim their freeholds and attract mates. The Bittern's foghorn-like call, for example, acts as a sound beacon in the reed bed.

During the breeding season, most song birds actively re-establish the ownership of their territories by singing boisterously at sunrise and sunset to produce the phenomenon of dawn and evening choruses. Purely visual displays may serve the same purpose, particularly for large birds of the open, such as Great Bustards, which adopt incredible postures on the open plains or grassy steppes where they breed. More usually, birds use a combination of visual and acoustic signals as in the display flights of Lapwings or Skylarks, or the bizarre posturing and cooing of Black Grouse on their 'leks' (see page 98).

The Tawny Owl (*above opposite*), a nocturnal woodland species, is the most vociferous bird found in Europe. The Great Bustard (*below opposite*) has another way of advertising its presence; it adopts a strange posture on open ground. The Black Grouse (*right*) will posture in addition to making a noise.

Courtship as a Language

Rapport is established between cock and hen by an exchange of signals, each with a specific meaning. Although the details of this language of love will vary from one species to another, the discreteness of the displays can best be illustrated by examining a species such as the North Atlantic Gannet. Life in a gannet colony situated on fairly steep cliffs is very competitive and so these birds are accordingly aggressive and assertive; timidity is no virtue and would mean failure to procure a nest site. But, because of the keyed-up hostility of the males towards intruders, courtship has its difficulties.

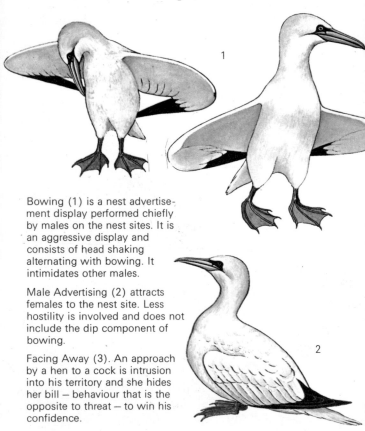

Bowing (1) is a nest advertisement display performed chiefly by males on the nest sites. It is an aggressive display and consists of head shaking alternating with bowing. It intimidates other males.

Male Advertising (2) attracts females to the nest site. Less hostility is involved and does not include the dip component of bowing.

Facing Away (3). An approach by a hen to a cock is intrusion into his territory and she hides her bill — behaviour that is the opposite to threat — to win his confidence.

Mutual Fencing (4) is a complex display involving dipping (chiefly by the cock) and bill scissoring that strengthens the pair bond, reducing any fear or aggression between the birds.

Sky Pointing (5) is a pre-departure signal that prevents the partner from also flying away and leaving the nest unguarded.

Understanding the Courtship Ceremonies of Birds

The elaborate rituals that birds perform during their courtship period have evolved over the ages in just the same way as structural features. We accept the fact that the wing of a bat or the flipper of a whale have developed slowly from a fore-limb of the kind more characteristic of terrestrial mammals; behaviour patterns, particularly those which are largely inborn, also have had an evolutionary history. As the bat's wing was evolved from a more orthodox terrestrial mammal's forelimb, so the complex displays like those described for the Gannet have been moulded out of simpler kinds of behaviour.

About sixty years ago, Julian Huxley (now Sir Julian Huxley) described the highly elaborate and beautiful displays of the Great Crested Grebe as *rituals*, and the evolutionary process by which simple forms of behaviour patterns are formalized into displays is now called *ritualization*.

Some displays of the Great Crested Grebe described by Huxley. The cat display (*below*) is a defensive attitude adopted by a bird that is sexually motivated. The penguin dance (*opposite*) is one of the strangest courtship dances observed among birds.

Ritualization

The process of ritualization is concerned with improving the impact of the signals and making their meaning more effective. After all, displays do form a language of love and rivalry and it is in the interests of communication that the ceremonies should broadcast their information clearly and unambiguously. The same holds true of our own society where important

signals with the same meaning tend towards the same form so that they are instantly recognizable; red lights whether on traffic or railway signals mean danger – stop, and against a background of noise, a telephone bell with its typical rhythmic pulses of sound can always be picked out immediately, even if there are other bells ringing. Bird displays often tend to be rhythmic, as in the bowing of pigeons or the bouncing dances of Australian Gouldian Finches, and these ceremonies usually have their own rhythm, which is characteristic of the species, like the telephone bell. Not only does the rhythm help to make the meaning of the display clear but also it tends to make the displays more conspicuous. A single male Gannet performing his nest advertisement display draws attention to himself among a mass of incubating and courting neighbours because of the formal and typical rhythm of his movements, which bear little resemblance to other Gannet ceremonies.

Of course, other methods are used to give the movements more impact. Certain components of the displays might

Mutual head shaking is performed by mated birds; it may start spontaneously or be induced by the sight of other pairs in full display.

become very exaggerated (like the rotation and spreading of the wings in the cat attitude of the Great Crested Grebe). The visual impression of the postures or rituals may be further improved by the addition of bright colours to certain parts of the plumage that are revealed to their best advantage during the ritual. There is every reason to believe that the ceremonies evolve *first* and the colour patterns are added later as a kind of garnish to make the displays more effective. Structural alterations of the plumage might even be made to enhance the courtship movements of some species. The aerial acrobatics of whydahs and drongos have become more eye-catching by the evolution of plume-like tail feathers that draw a potential mate's attention to the displaying birds from a much greater distance than would otherwise be possible. In our own society, for example, the frequent use of social gestures made with the arms and hands were, in days gone by, made all the more conspicuous by the development of cuffs or by holding pieces of lace. These techniques apply also to bird displays.

The hostile threat display is given to a rival.

79

The Raw Material of Displays

What is the origin of display movements? In our own species, some primitive dances can be looked upon as ritualized enactments of the hunter pursuing his prey; even eating and drinking have been incorporated into the social ritual of the cocktail party or formal dinner when they provide a common activity and help to 'break the ice'. A close study of bird

Greeting ceremonies are common to many wild-fowl when friendly birds go through the motions of drinking. These are Mandarin Ducks.

ceremonies shows that they too are often stylized forms of everyday behaviour such as drinking, preening or flight intention movements.

Flight Intention Movements

The preparatory phases of activities (or intention movements) have been used a great deal as building blocks of displays. The reason for this is that when birds display they are generally signalling their readiness to do something and this will involve movement. Courtship situations are also often tense and the emotions of fear, aggression and sex may be aroused simultaneously and pull in different directions. Conflict like this is likely to elicit *locomotory intention movements* as the bird prepares to *approach* the mate in order to attack or copulate with it, or to *escape*.

The preparations for flight involve two basic postures and one or other of these have become ritualized in many species.

Two phases of the take-off leap.
Firstly the bird coils itself up (1).
Then, like a spring, it uncoils and
launches itself into the air (2).

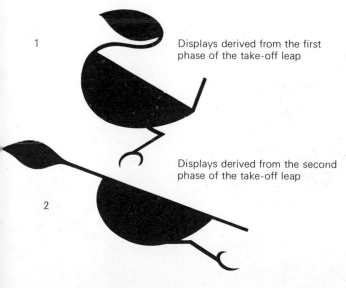

Displays derived from the first
phase of the take-off leap

Displays derived from the second
phase of the take-off leap

1. A Mallard in the head-up tail-up posture, just one of the many postures that make up a courtship sequence.

2. A Mute Swan in threat

3. A Cormorant's rhythmic wing-flapping exposes the white thigh patch.

4 and 5. The Hooded Merganser has courtship postures derived from both phases of the take-off leap.

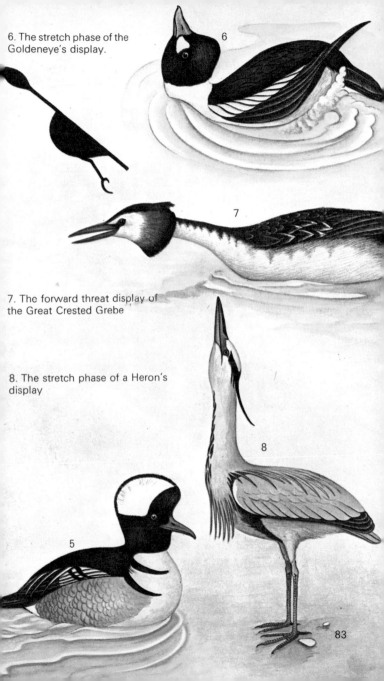

6. The stretch phase of the Goldeneye's display.

7. The forward threat display of the Great Crested Grebe

8. The stretch phase of a Heron's display

83

Courtship Dances

Sometimes a whole series of take-off leaps have evolved into rhythmic dances. During the second phase of its courtship, the male Gouldian Finch jerks its body up and down in front of the hen. While performing the ritual, the feathers of the nape are ruffled, framing the head in cobalt blue, and those of the breast are raised to make the lilac patch on the chest larger. The majority of Gouldian Finches have black heads, although about one in four has a red crown and cheeks; there is a rare yellow (or orange) headed variety. The normal posture and courtship dance of the red-headed birds is illustrated below.

Dancing may take the form of pivoting, reflecting the

underlying conflict of the cock between approaching in order to mate and avoidance. In the Zebra Finch, every swing turns the cock away from the hen, then brings him back towards her again. With each pivot, he hops along the branch in her direction but then swings away from her again. This swinging towards and away from the hen signifies the underlying conflict between approach and avoidance that characterizes many courtship rituals of animals.

Cranes are well known for their crazy courtship dances, and their leaping and prancing behaviour might be derived from flight intention movements. These dances occur after the cranes have migrated north for the summer. When the flocks arrive, they divide up into pairs and the courtship dances commence. The male moves towards the female, bowing with its neck bent forward; he then leaps up into the air several times, making trumpeting calls. The female joins enthusiastically in the dance, splashing through the shallow reaches of water where the birds perform these dances.

Crowned Cranes dancing

Displacement Activities

In conflict situations, when quite different moods such as fear and sex are aroused, birds may show behaviour which seems to be entirely inappropriate to the circumstances. If faced with an aggressive hen, a sexually-aroused cock might preen, wipe its beak, or scratch. This is the avian equivalent to someone head scratching or foot tapping when faced with an apparently insoluble problem. 'Irrelevant' actions of this kind have been categorized as *displacement activities*, and in many species they have become ritualized and incorporated into displays.

Beak-wiping is a common displacement activity. The Spice Finch has a special courtship posture called the *low twist* which is really an incomplete and frozen beak wipe; in courtship situations the bill is never actually rubbed on the perch (*top right*).

Birds often preen when they are 'nervous', and these nervous or displacement responses then develop into displays. The Shelduck (*top opposite*) vigorously preens its wing when courting but in the exquisite Mandarin drake (*bottom opposite*), the preening is merely a symbolic touching of the sail-like wing feather.

Lovebirds are often very hostile to their mates and head scratching is frequently shown by them. In the Madagascar Lovebird (*bottom right*), the scratching is almost normal whereas in the Black-masked Lovebird (*centre right*) it is ritualized, the foot being frequently raised to the red bill; in fact this ritual might even serve to draw attention to the brightly coloured display structure.

Displays for Calming the Mate

A number of aggression-reducing (appeasement) and fear-reducing (reassurance) gestures are usually built into the courtship language of birds so that these disrupting emotions can be held in check to allow the business of breeding to go ahead smoothly. A bird's chief weapon is its bill and during threat this is pointed at the opponent; most appeasement displays involve pointing it *away* from the mate.

Mutual facing away appears early on in the pair formation process of gulls, when both partners are not yet fully used to each other and may be seized with an impulse to attack or flee from each other. In these Herring Gulls, facing away or head flagging is an important appeasement ceremony.

Displays Derived from Redirected Attacks

An angry bird which is prevented from attacking a hated rival may vent its rage on another individual (a third party) or on something entirely different. As shown below, a Herring Gull, confronted by another over a territorial boundary, will furiously rip up tussocks of grass; this behaviour is ritualized to some extent, serving to warn the opponent: 'This could happen to you – so leave me alone!'

Ritualized Nest-Building

As courtship and nest-building occur so closely together in time, it is not surprising that some of the specialized movements that are used to incorporate material into the nest have been themselves modified and worked into the breeding ceremonies. The mutual penguin dance of Great Crested Grebes is carried out with pieces of weed in their bills, although here any similarity to nest-building ends (see page 77). Herons and Black Skimmers will pass nest material to the mate at some stage during the courtship.

In the Adelie Penguin rookeries of Antarctica, a penguin will offer its mate a pebble, a valuable aquisition in the snowy wastes. The pebble is more often than not stolen from a neighbour's nest pile (*below*).

Ritualized nest-building movements are seen in their most sophisticated form in the straw displays of African waxbills. These birds build domed nests using both upward-stretching

and tucking beneath movements in the building process. It is the former motor pattern that has been ritualized in the case of the African species, although oddly the Asian related Red Avadavat uses the latter in a modified form in its bowing display.

The stem display of the Blue-breasted Waxbill showing the highest and lowest positions reached in the rhythmic 'dance' (after Derek Goodwin).

The bowing display of the Red Avadavat is derived from a tucking-in nest-building movement.

Blushing and Ruffling

Sudden sweating or flushing of the skin with blood, brought on by being faced with an uncomfortable or conflicting situation, are responses with which we are all familiar. They are just two of a number of adjustments brought about by an arousal of the *autonomic nervous system*; briefly this system prepares the body to 'fight or flee'.

When their emotions are fired, birds show a number of responses that suggest a strong arousal of the autonomic nervous system. These responses include a general erection of the feathers, which is the avian equivalent to sweating because it allows more heat to escape from the body, and contraction of the pupils. Both are characteristic of bird displays, and species such as turkeys posture to their mates and rivals with every single feather raised to the full.

Although a basic arousal of the autonomic nervous system might result in a complete ruffling of the plumage, evolutionary changes have brought about a *restriction* of feather erection to certain special sites to make the signals more effective. For example, in the grassfinches, the cock Red Avadavat displays with most of his plumage ruffled (the primitive condition) whereas the African Lavender Finch raises its brilliant red rump feathers most of all; similarly, the Cut-throat Finch has a most distinctive pattern of feather erection.

A Lavender Finch in display showing the special feather erection on the rump.

In large birds such as turkeys, blushing is another autonomic response that has been developed for display purposes; the bird's mood is accurately recorded by the changing blue and red hues of the naked areas of skin on the head and neck.

A cock Cut-throat relaxed (*left*) and displaying (*centre*) to show the pattern of feather erection and the way this helps to display the special plumage markings to the hen (*right*).

Further Evolutionary Changes

We have already seen how the basic autonomic response of feather erection might be modified by restricting the area of plumage ruffling. The impact of the signals can be further enhanced either by developing colours that can be dramatically exposed when the plumage is ruffled, or by modifying the

The beard of the cock Capercaillie (*top*). All kinds of birds have crests. Leadbeater's Cockatoo (*centre*) exposes a dramatic pattern of red and pink when its crest is erected; the Blue-crowned Pigeon's is permanently erected (*bottom*).

Two of the most spectacular bird displays are those of the Peacock and male Argus Pheasant (*opposite*). Both produce great fans of feathers, one using the upper tail coverts and the other the wings. They have some of the longest of bird feathers, beautifully adorned with eye-spots of iridescent colours. The Argus Pheasant's wings, modified as they are for display, make it difficult for the bird to fly.

peacock

Malay Argus Pheasant

feathers that are erected, or by both. Over the course of evolution, crests, beards, tufts, plumes, trains and so on have become established independently in various families of birds, and these are related in some way to the methods of courtship employed; the courtship movements and postures come first and these at a later evolutionary stage become garnished with colours and modified feathers. For example, many pheasants that present themselves frontally to their mates have evolved displays that are optically most effective when seen from the front (e.g. the Peacock and Argus Pheasant) whereas those that display laterally, such as the Golden or Lady Amherst's Pheasant, have capes of feathers around the neck that can be fanned and look most dramatic when viewed broadside on.

Acting Young

Some courtship rituals have obviously originated from juvenile behaviour. Courtship feeding is one such ritual, and is known to occur in no less than fourteen orders of birds, including seventeen families of Passeriformes. Usually, the hen begs food from her mate in a manner characteristic of the young of the species soliciting food from the parents. Of course, in species where the hens do most of the incubating (Great Tits) or all of the incubating (hornbills, where the hen is entombed in the nest cavity), this 'free' food is vital for their survival, besides serving to cement the pair bond. Its precise nutritional value in most other species has yet to be investigated.

Courtship feeding is a feature of the precopulatory behaviour of many finches, including Goldfinches. The female is shown on the nest.

The begging posture of two young Red Avadavats (*above*) and the pair formation display of the mature cock (*right*). An adult hen is shown on the right in both cases.

Cock Red Avadavats have a pair formation display which bears an astonishing resemblance to the specialized juvenile begging postures of the species. The young turn their heads towards the parent, raise, and quiver their off-side wings.

What is the significance of these juvenile types of behaviour? Both the food soliciting by hens and the wing-raised posture of the cock avadavat might help to arouse parental responses in the sexual partner and consequently counteract moods like fear or aggression that would operate to disrupt the pair bond. Of course, they are now highly stylized forms of display and they may not produce real parental responses, but at least this explanation gives a clue as to how they evolved in the first place.

Arena Displays

For the majority of birds, pair formation is the central event of their breeding season. However, no lasting pair bond is formed in ninety-two species; the sexes live separately and come together for only a few seconds or minutes each breeding season in order to copulate. How is this union achieved?

The cocks often assemble on traditional display grounds called *arenas* or *leks*, where each bird defends a small plot called a *court*. The sizes of these arenas vary from one species to another; for example, the Greater Prairie Chicken – a grouse from North America – displays in arenas up to 200 yards wide and half a mile long holding the courts of 400 cocks. Individual courts may be dispersed, as with the Argus pheasant, or very concentrated, as in the case of the Blue-backed Manakin or Ruff. Nearly all cock 'lek' birds have splendid plumage and this is shown off on the arenas to attract the hens. However, the cock birds spend as much time displaying to each other in order to maintain their status within this temporary all-male society, and much of it is ritualized fighting.

Two cock Prairie Chickens displaying, and a hen

European Ruffs hold their arenas on grassy meadows. The hens are small and dull by comparison with the cocks which have wattles, ear tufts and ruffs, and these vary in colour from black through brown, red or yellow to white. No two cocks on a lek are similar and so individual recognition is easy. However, this lek species is unusual because of behavioural differences between the birds that are related to the role they play in courtship. *Resident* cocks hold their own courts, tend to be aggressive, have dark coloured ruffs and tend to mate more than *satellite* males that do not possess their own courts; these males become associated with resident cocks, are less assertive and have white ruffs serving to draw the hen's attention to the arena and their own resident cock. On a large lek the residents have a monopoly of copulation, whereas on the smaller ones the presence of a satellite male increases the chances of copulation for both residents and satellites.

Hens visit the leks and they solicit on the residence of their choice male; being promiscuous they may fly from lek to lek before retiring to lay their eggs

European Ruffs displaying

Cock-of-the-Rocks are arena
birds. The colourful males take
no part in rearing the young.

The Superb Lyre Bird (*below opposite*), one of the two species of lyre birds in the thick forests of Australia, is a spectacular arena bird. Each cock prepares his own courts or clearings (up to ten in number) in the undergrowth into which the hens will be enticed. At the climax of the display, the highly ornate, silvery tail is brought over the head like a 'fairy parasol' – a sight of rare and moving beauty.

Birds of paradise from the New Guinea region must be among the most ornamented birds and at least twenty-four of them display in arenas. Three are shown displaying below.

1. Magnificent Riflebird
2. King of Saxony's Bird of Paradise
3. Magnificent Bird of Paradise

Bower Birds

Several species of birds have been recorded as modifying their surroundings to make their own displays more effective; for example, the Magnificent Bird of Paradise trims the leaves around its court, so allowing more light to reach its glistening metallic plumage. Bower birds are even more remarkable because the males build structures like tents, and tend 'gardens' which they decorate with bright berries or even jewellery if they can find it, to impress and attract the

females. It does seem that these artifacts have come to replace the plumage as a form of display; the males have become relatively dull, with concealing coloration, but make up for it by constructing eye-catching bowers. There may be considerable rivalry between neighbouring cocks who may raid each other's bowers and steal bright trinkets for their own.

Many bower-birds, such as the Satin Bower Bird (*right*), are avenue builders and the hen has to be enticed inside the avenue before mating takes place. This species decorates its bower with blue objects. The cock is shown in the foreground with the hen behind.

The Stage-Maker Bower Bird clears a court and decorates it with leaves of selected species of trees which are sawn off with his serrated bill and placed pale-side up on the ground (*right*). If turned over, then the cock will replace them, pale-side up. He does not perform any visual display.

On the whole, the duller the plumage of the males, the more elaborate the bowers. The Crestless Gardener builds an incredible tent-like structure surrounded by a carefully attended garden, turfed with moss and decorated with flower petals. These are renewed almost daily (*opposite*).

Song

Songs are part of the complex communication system that comes into operation during the breeding season. Although on the whole we find them pleasing to listen to, they are really two-edged weapons and their information content depends on who the listener is! As it is the males who sing, the sound signifies the sex of the performer and advertises the fact that he has a territory and may be unmated. Hens may be attracted, and are undoubtedly stimulated, by the song; hen Budgerigars

The Chiff-Chaff and Willow Warbler are almost identical and might occur in similar habitats; yet confusion and interbreeding is prevented by the development of different songs, as shown by the spectrograms below.

Chiff-Chaff

frequency

time

Willow Warbler

frequency

time

Distribution of the Chiff-Chaff

Distribution of the Willow Warbler

kept within earshot but out of sight of warbling cocks ovulate sooner than completely isolated birds. Nevertheless, to other cocks of the same species, a singing bird is broadcasting the message 'keep out'.

Every species has its own special song for attracting a mate which differs from all other closely related forms. Hens finally have to choose a mate and it is as though they are attracted by their own species-specific songs. Sometimes the song is most important in preventing hens making the wrong choice.

Eastern Meadowlarks and Western Meadowlarks look remarkably similar and occur together in places, yet no interbreeding occurs. The spectrograms show song to be the main isolating mechanism.

Eastern Meadowlark

frequency

time

Western Meadowlark

frequency

time

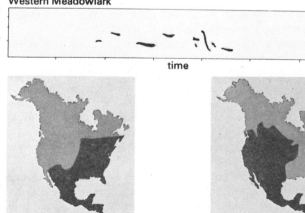

Distribution of the Eastern Meadowlark

Distribution of the Western Meadowlark

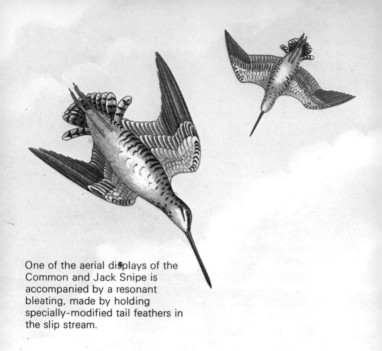

One of the aerial displays of the Common and Jack Snipe is accompanied by a resonant bleating, made by holding specially-modified tail feathers in the slip stream.

Songs Without Voices

In widely separated groups of birds, the chief form of advertising is by instrumental means and these are often refinements of more ordinary activities that make sound.

For example, flying itself creates a certain amount of noise and, in many birds, the wings are used as sound-producing organs to the extent that the feathers may be modified to achieve this end. Wood Pigeons clap their wings in a resounding way during their display flights, and the Ruffed Grouse produces a machine-gun like rattle by beating its wings while standing in its territory, a noise that carries at least half a mile through the forest habitat. Tail feathers are used by the Sharp-tailed Grouse, which rattles its quills.

The first primary wing feather of the Broad-tailed Hummingbird is altered to enhance the humming noise. These birds perform display flights consisting of spectacular acrobatics, and, during the dives, the wings beat at 200 beats per second

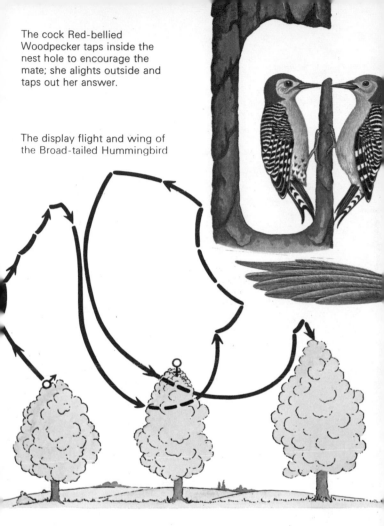

The cock Red-bellied Woodpecker taps inside the nest hole to encourage the mate; she alights outside and taps out her answer.

The display flight and wing of the Broad-tailed Hummingbird

to produce a rattling whistle as air passes through the special slots between the first and second wing feathers.

Mechanical signalling reaches its highest form in some of the woodpeckers which rapidly strike their bills against branches. Each species has its own recognizable tattoo, differing from others in duration, number of taps, intensity or rhythm.

Songs Designed for Survival

Like structure and plumage coloration, bird songs have been exposed to the rigours of natural selection so that, for any given species, the duration, pitch, rhythm and tonal qualities are adapted to the conditions in which they have to operate.

An advertising bird is a vulnerable one because it is drawing attention to itself by *visual* or *acoustic* means; such is the danger from predators that only rarely are both methods used at the same time. A highly camouflaged species, like the Song Thrush, may sing loudly and often, but brightly coloured birds tend to have lesser songs or to sing discontinuously.

Birds that live in thick vegetation may be protected from enemies and are not easily seen by a potential mate; their songs are often loud and continuous, acting as sound beacons. Reed-inhabiting warblers and marsh wrens are examples of these birds. They may also

The Nightingale, Nightjar and Savi's Warbler (*top* to *bottom*) are all protected from enemies to some extent. They have continuous songs heard at night.

The Blackbird (*above right*) has one of the most musical 'thrush'-

sing at night like the Nightingale so-loved by the poets. Species such as nightjars and whip-poor-wills find concealment in the darkness and they also have continuous, monotonous vocalizations. Birds that forage in the safety of the canopy, such as tits, may sing during feeding, but those that feed on the ground where they are vulnerable hardly ever do so, such as thrushes. These birds may often be highly camouflaged and consequently have well-developed songs of good carrying power, delivered from tree perches.

The tonal qualities of the song have an obvious bearing upon its carrying power. Low-frequency notes carry a long way, particularly in dense habitats like woods and reed beds. It is, therefore, significant that bitterns have powerful songs sounding like fog-horns; owls that live in woods and are nocturnal, are more vocal and have deeper voices than diurnal or crepuscular species of similar size.

type songs; it sings close to cover and is not too conspicuous. Its relation, the Ring Ouzel (*right*), inhabits moorland and is a conspicuous bird — it has one of the crudest 'thrush'-type songs.

Duetting

Among various groups of birds that are more or less restricted to the tropics, there are species that are capable of duetting. The notes of the cock and hen are different and the sexes alternate their parts to sing *antiphonally* in a very precise way so that, to the untrained and uninformed ear, the performance sounds like the utterance of a single bird.

For example, the Magpie Lark of Australia sings antiphonally; one bird calls *'te-he'* and the other quickly replies *'pee-o-wit'*. The Marbled Wood Quail of the Central American forests whistle at dusk with a *'oo-oo-oo-*oo*-oo'* repeated many times, the last two syllables being uttered by the mate.

Duetting reaches its highest development in African shrikes of the genus *Laniarius*, commonly known as bou-bou shrikes. The songs of these birds are tonally very pure and musically pleasing and their powers of invention well-developed. For

A pair of bou-bou shrikes

example, a pair of these birds may have a very extensive repertoire of duetting patterns, from very simple ones to more involved antiphonal melodies. Some of these have been set out below.

It seems that these duets are gradually worked out by the members of a pair during extensive 'rehearsals', that either sex can start or finish, and that either bird can sing the *whole* pattern alone if the partner is absent.

What is the use of duetting? Firstly, it tends to be found in those species that inhabit thick vegetation and so might be important in keeping the two birds in contact. The bird that responds with the correct phrase, therefore completing the duetting pattern, is the mate; this is all the more likely as each pair has duetting patterns that they do not share with other neighbouring pairs, although, of course, some of the simpler patterns may be common and have a wide distribution.

Six duetting patterns of a single pair of bou-bou shrikes. The contributions of each bird are set out in different colours, red indicating the first singer (after Thorpe).

Learning the Language

It might be thought that birds would automatically reproduce their own songs during the breeding season. Although this is true of some species, it has recently been found that experience and learning play a much greater role than had previously been suspected.

Doing What Comes Naturally! It is probable that brood parasites like the Cuckoo are able to develop their full vocal repertoire without picking up clues from more mature members of their own species; indeed this is particularly important insofar as they are exposed during their early life to the vocalizations of their foster parents. Their language might, therefore, be genetically fixed. Even strongly migratory species such as the Common Whitethroat appear to develop the species-specific song automatically.

The Role of Learning The song of the Chaffinch has been studied more than any other species and the findings are most revealing, although they have to be applied cautiously to other species. The normal song of the territory-holding male consists of three phrases. The first descends the scale but increases in

Spectrogram of song of normal Chaffinch

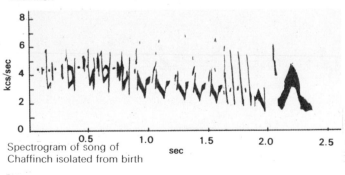

Spectrogram of song of Chaffinch isolated from birth

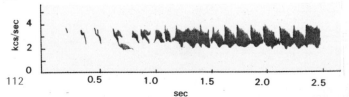

Cock Chaffinch singing

volume and is followed by a series of two to eight notes which are very similar; the terminal phrase ends with a flourish.

Young individuals caught when only a few months old were still able to develop a more-or-less normal song pattern providing they were in each other's company. However, cock Chaffinches hand-reared from the egg and kept out of hearing of other Chaffinches produced completely abnormal songs; they were of poor tonal quality, although of approximately the right length ($2\frac{1}{2}$ seconds) with the right number of notes, but they never learned to phrase these notes correctly. This is the *innate* contribution to the Chaffinch's song.

It seems the cock Chaffinch goes through two sensitive periods when it picks up details from other fully-grown birds to embellish and develop its crude inborn pattern. The first period is when it is in, or has just left, the nest before it can sing; the other is during the following spring when the song becomes finally fixed in its mind after counter-singing with other Chaffinches living in the same area.

The young cock Chaffinch must be exposed to the songs of other cocks of the same species in order to develop its proper song. However, the precise effect of this kind of experience varies in other species.

The young White-crowned Sparrow learns its song during the first hundred days of life before it can sing and so stores up the information to use later on; as with the Chaffinch, failure to gain this experience results in the production of an aberrant song. If adult songs are played to Arizona Juncos, the young cocks develop songs characteristic of the species *but* these songs are unlike any of the teacher songs; hearing the songs stimulates the birds' vocal inventiveness and prevents them from copying.

Quite a different situation prevails in the Oregon Junco. Deprived birds tend to be very inventive and seem capable of learning elaborate song patterns. However, the song of this species is a simple trill, and exposure to mature adult cocks' song causes them to curtail their inventiveness.

Oddly then, the Oregon Junco which has a simple song has to learn it whereas the related Song Sparrow which has a much

Cock Oregon Junco

more varied and tuneful song does not have to pick it up by cultural transmission.

Which Song to Copy? This is a puzzle and it can only be assumed that birds tend to copy those songs and sounds that approximate most closely to their own innate pattern of vocalizations; in other words, they may pick up song patterns that they can reproduce most easily.

Song Dialects? The songs of male Cuckoos or Whitethroats are much the same no matter from what part of the range they are heard. However, species such as the Chaffinch develop quite marked local dialects, and these are handed on culturally to young birds which tend to breed in the area where they are born, handing on the dialect to their own young. The obvious parallel can be seen in the human race.

Birds with adjoining territories tend to challenge each other by song and there is some evidence to show that songs in the local dialect produce a much greater response than those with a more alien dialect, at least with Red Cardinals.

Cock White-crowned Sparrow

Mimicry

As the ability to copy phrases of song plays such a prominent role in the life of birds, it is to be expected that some species plagiarize the languages of others in a wholesale manner in constructing their own songs. Old World species such as the Marsh Warbler and Starling are well known mimics; one individual has been known to reproduce snatches of the vocabularies of over thirty species. One of the most famous mimic species is the Australian lyre bird which is a veritable one-man band. In Africa there are the robin chats, and North America's Mockingbird ranks as one of the world's best singers, although some of its phrases are those of other species.

Talking Birds

Species that are accomplished mimics, like the parrots and mynahs, can produce plausible imitations of human speech if they are kept under confined conditions; one Amazon Parrot could enunciate between fifty and a hundred words and this is no record. These birds probably form a social attachment to their human keepers and learn that 'talking' on their part is rewarded by social contact; they often talk most volubly when alone, perhaps attempting to bring their owners back.

Cock Marsh Warbler

Cock robin chat

The question arises as to why some birds imitate sounds. Imitation does result in a more varied song, although thrushes achieve the same ends by rearranging their song phrases into different orders. This means that each individual might have a song which is distinguishable from all other members of the same species, although still recognizably characteristic of the species. It makes it easier for mates to recognize each other by their vocal utterances. However, the best mimics of the human voice do not seem to use their imitating ability fully in the wild.

Two of the best talkers are the Indian Hill Mynah (*bottom*) and the African Grey Parrot. The spectrograms are of the sentence 'You make me laugh', spoken by a human and an Indian Hill Mynah, and show the functional similarity.

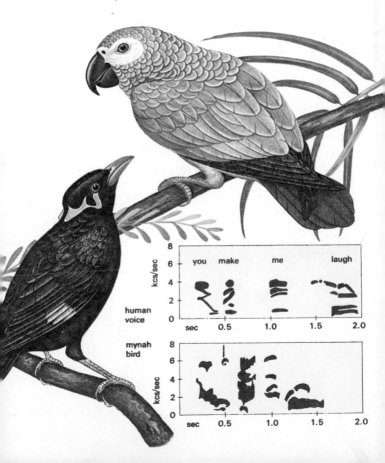

Nest Building
Who Builds?

The extent to which the sexes co-operate in bringing material to the nest site and in building the nest varies tremendously from family to family. In kingfishers, both birds help to excavate the tunnel, whereas in weaverbirds and grassfinches the males do most of the work. The hens are the home makers in the manakins, vireos, hummingbirds and finches.

The Mechanism of Nest Building

Whether males, females or both sexes indulge in nest building, it involves a lot of time and industry; for example, over 1,200 mud-carrying trips were made by a pair of Barn Swallows before their flask-shaped abode was finished, and a nest of a Black-throated Oriole was found to contain 3,387 separate

pieces of material. This burst of constructive energy must be well timed in relation to other events in the breeding cycle, such as egg-laying; it does not happen automatically. Experimental analysis has shown that this behaviour is triggered off, controlled and brought to an end by a combination of interacting *internal* and *external* stimuli.

Hen canaries build the nest in two phases. Firstly the outer cup is constructed of coarse grass or twigs and then feathers are used to line the cavity just before the eggs are laid. Nesting behaviour is stimulated by female sex hormone (oestrogen), which is produced under both the influence of increasing hours of daylight in the spring and by the courtship behaviour of the cocks, which intensifies at this time of the year. The nest is, therefore, started but as building proceeds, stimulation from the actual nest influences the hen's choice of nest material, but in an indirect way. With increasing levels of oestrogen in the blood, the hen commences to shed some breast feathers, leaving bare sensitive brood patches. These become highly vascular just before the eggs arrive and they will serve to keep the eggs warm. It can well be imagined that the rough-textured grass will stimulate these sensitive areas at this time, causing the hen canary to change her choice of nest material to feathers to cut down the sensation and also provide a soft, well-insulated interior to the nest cup.

3

1. The influence of increasing daylight and courtship by the cock causes the hen to build a nest of twigs or rough grass.
2. Further courtship and stimulation from the nest causes the hen to shed feathers over the brood patch, which thus becomes sensitive.
3. The sensitivity of the brood patch causes the hen to line the nest with soft feathers.

The Variety of Nests

There are almost as many kinds of nests as there are different sorts of birds. At one extreme, there are King and Emperor Penguins and ledge-nesting auks which have no fixed nest site and build no nest; the penguins incubate their single eggs between their shanks and bellies. At the other, there are the vast communal nests of Sociable Weavers; one incomplete structure (part had broken off) measured twenty-five feet by fifteen feet by five feet and contained ninety-five nests. Only the Australian megapodes' nests could exceed this size.

Basically, the nest is a place where the eggs are laid and young nurtured, providing them with security, support and insulation. Of course, the

The King Penguin builds no nest but incubates its egg beneath a fold of skin.

The Nuthatch nests in tree cavities and then plasters up the entrance just enough to exclude larger species like Starlings.

The open cup is a common type of nest. The Chaffinch weaves in moss and lichens on the outside as camouflage.

site and presence of enemies may influence the form of the nest. Secure on their ledges, guillemots build no nest and lay pear-shaped eggs that tend to spin round on the spot; on the other hand, many tropical perching birds build finely-woven structures, often hanging over water, with stocking-like entrances that help to deter snakes.

All manner of materials may be used from saliva (as in cave swiftlets), mud (Cliff Swallows), fibres, stones, feathers (Eider Duck) and even snake skins (crested flycatchers).

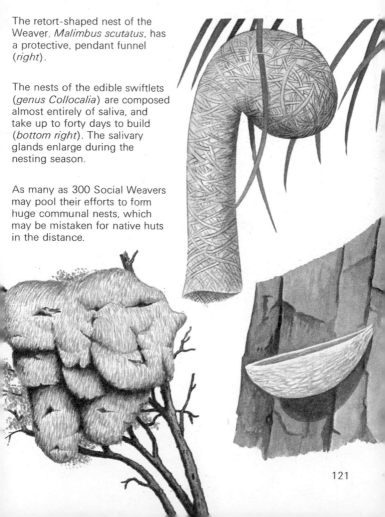

The retort-shaped nest of the Weaver, *Malimbus scutatus*, has a protective, pendant funnel (*right*).

The nests of the edible swiftlets (*genus Collocalia*) are composed almost entirely of saliva, and take up to forty days to build (*bottom right*). The salivary glands enlarge during the nesting season.

As many as 300 Social Weavers may pool their efforts to form huge communal nests, which may be mistaken for native huts in the distance.

By far the majority of birds use vegetable material such as grass for building. One remarkable species, the Tailor Bird, sows a number of leaves together, first piercing their margins with its thin pointed bill and then binding them with fibres knotted at one end to prevent them pulling through. The nest is then built up inside the cup formed by the leaves.

Birds that use fibrous material employ a relatively small number of stereotyped movements to weave, tuck, and thread the material in order to form the structure of the nest. Among these birds, the weavers appear to be the most skilful in the way they build up the fabric, and some build nests of advanced design that appear to be constructed according to a well-formed plan. Basically, each nest consists of a basket with two

A Tailor Bird and its nest

chambers and a roof lining, and a layer of fine material is placed around the nest chamber.

Although weaverbirds build nests of varying design, certain common features follow from the fact that the nest basket is built up largely from a fixed position on an initial ring and that the dimensions of the nest and the antechamber are chiefly determined by the reach of the cock. Here it can be seen how the cock Village Weaver builds up his nest. It is from the initial ring that the male hangs upside-down advertising for a mate.

The Village Weaver builds its nest of palm fibres. First it makes a ring (1) and enlarges it step-by-step (2) into an egg chamber (3). It then constructs an antechamber (4) and finally an entrance to the nest (5).

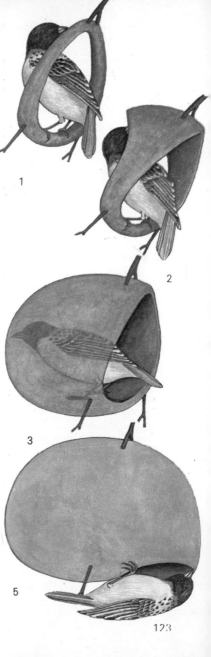

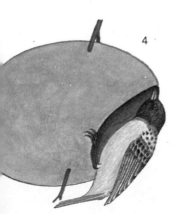

Incubation

The eggs must be kept at a relatively high temperature or else the embryos will fail to develop. Under the influence of hormones, most birds produce brood patches to facilitate the transfer of heat to the clutch. Gannets and cormorants use their highly vascular feet for the purpose. The longest recorded incubation period is eighty days for the Royal Albatross; many of the smaller perching birds take only ten days to hatch their eggs.

Who Broods? All variations in brooding behaviour exist between the rarer system of 'males only' (Emperor Penguin, Kiwi, Ostrich, rheas) and the method of hornbills, in which the hens are sealed into the nest cavity by their mates who then feed them through a narrow aperture left open especially for the purpose. In those species in which the males are brightly coloured, their presence near the nest may jeopardize it and the family and so the more camouflaged hens take over. However, in well over half of the families of birds, both sexes share the task of incubation.

Egg Retrieval Ground-nesting birds like gulls and geese will often attempt to retrieve eggs that become accidentally displaced. A Greylag Goose does so by stretching out its neck, placing its bill on the other side of the egg and slowly guiding it back into the nest cup; of course, it compensates for any tendency of the egg to roll in other directions, although rather clumsily. However, arboreal nesting species that have deeply-cupped nests do not show this behaviour at all. This lack of

The Whitethroat is one of several birds that show no concern for eggs displaced from the nest.

The under surface of a female Black-headed Gull showing how the eggs fit into the brood patches.

concern for a displaced egg or even a displaced chick leads people to believe that the birds are heartless or stupid. In fact retrieving behaviour has no relevance in a situation where a displaced egg inevitably means a broken one on the ground.

Slaves to the Mounds Not all birds use the heat from their bodies to hatch their eggs. Some of a group of twelve Australian and Malayan megapodes (brush turkeys and incubator birds), which are rather like pheasants, build mounds up to ten feet high around their egg chambers. Hens visit the mound to lay the eggs, which are then covered by rotting vegetation and soil by the strong-footed cock; the mounds are really nothing more than compost heaps. Heat from the sun and decomposing plant material is sufficient to allow the eggs to develop. However, the temperature inside the mound must be maintained at a more or less constant 92°F no matter what the weather is, and so the males must work with unremitting energy for up to thirteen hours each day to adjust the dimensions of their mounds according to the heat balance; failure to do so will result in the eggs becoming chilled or cooked! Should the egg chamber within the rotting compost become too hot (the temperature can easily reach 113°F), then earth is removed from the top in order to allow heat to escape; the mound is restored to its original condition at night when the air temperature drops and there is need for insulation.

Of course, cock megapodes must be able to determine the temperature of their nests so that they can take the appropriate

A Mallee Fowl and its mound shown in cross-section in order to illustrate the structure. When more heat is required, the cock builds up the insulation layer in a thick layer over the compost. During the night, the heat may

action; this they do by taking some of the soil or compost in the bill and pressing it against their heat-sensitive palate or tongue. This is done regularly.

The Australian zoologist, H. J. Frith, has studied the responses of the Mallee Fowl cock to changes in the temperature of the mound induced by placing an electrical heating element within the egg cavity. When the heater was turned on, the cock would remove some of the soil to allow the heat to dissipate, thereby adjusting the temperature.

Apart from the mound builders, the Maleo Fowl from the Celebes lays its eggs singly into black volcanic sand, which absorbs the heat of the sun very well and thus keeps the eggs developing. Sometimes the eggs are laid in volcanically warmed soil where they are incubated by volcanic steam.

increase rather too much and so first thing in the morning, the cock Mallee Fowl gets to work scraping away the centre to allow heat to escape. The chicks have to fight their way to the surface when they hatch.

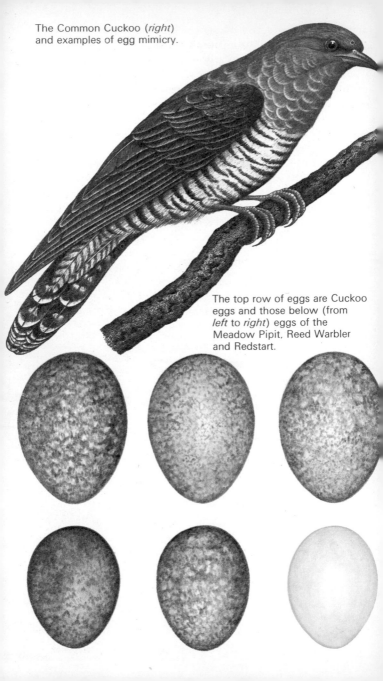

The Common Cuckoo (*right*) and examples of egg mimicry.

The top row of eggs are Cuckoo eggs and those below (from *left* to *right*) eggs of the Meadow Pipit, Reed Warbler and Redstart.

Brood Parasitism

Some species lay their eggs in the nests of other birds which then have the task of rearing their adopted offspring. Brood parasitism has evolved in the Black-headed Ducks, honey guides, cow birds, weaverbirds, viduine weavers and forty-seven species of cuckoos.

The habit of fostering out the young raises certain complications and also demands special behaviour from the parasites in order to ensure their offspring's survival. The host must feed on roughly similar food to that of the nest parasite and their egg sizes and incubation periods must correspond. The eggs must not be too dissimilar and, once hatched, the young intruder must present the right signals to prevent the foster parents from feeding their own young in preference.

The European Cuckoo is known to patronize over 120 species although each hen probably specializes in one species of host such as Reed Warblers or Meadow Pipits. Only one egg is laid in each nest and, to compensate, one of the host's is removed. Once hatched, young Cuckoos, like many other brood parasites, show patterns of behaviour designed to reduce competition from their foster brothers and sisters.

The young Cuckoo (*right*) ruthlessly removes competition from its hosts' offspring. When newly-hatched, it will heave out both eggs and young birds from the nest with its shoulders. The young Cuckoo also has a relatively enormous yellow mouth with great fleshy flanges on either side; when this is opened, as is the practice when young birds are hungry, it provides an irresistible stimulus to the foster parents to bring food. A single young Cuckoo can make up in this way for the whole brood that it displaces.

Whydahs and Waxbills In Africa, the eleven species of viduine whydahs have evolved very sophisticated relationships with their waxbill hosts; these have been elucidated in great detail by Dr Jürgen Nicolai.

Unlike many other brood parasites, hen viduines may lay two or three eggs in each nest and the young make no attempt to evict their foster nest mates and indeed may continue the association with their foster family for quite a time after fledging. Each viduine species or race parasitizes only one kind of waxbill; they are host-specific. This specialization has been forced on these brood parasites because parent waxbills will

cock

hen

The cock Pin-tailed Whydah calls the attention of the hen to a nest-building cock Violet-eared Waxbill by uttering a series of nest-calling notes similar to those of the host species. The similarity of these nest calls can be seen in the spectrograms.

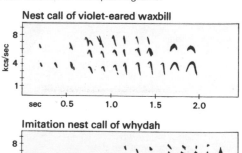

Nest call of violet-eared waxbill

Imitation nest call of whydah

Paradise Whydahs (shown in courtship postures, *above*) parasitize only the south and east races of the Melba Finch (*below*).

hen

cock

hen

only feed nestlings that show the correct oral display, begging behaviour and down pattern of their own species; an intruder will have to conform with these characteristics to survive. As a result, there is a high degree of mimicry by the parasite species of the host species in such features as mouth markings.

The fixation on the host species occurs during the early stages of the parasite's life. The cocks pick up and mimic with great fidelity the host's vocabulary and this vocabulary later plays a part in its own courtship; hen viduines prefer cocks whose songs contain a complete and accurate vocabulary of their joint hosts. This mimicking ability may also help the parasites to call up and therefore locate their hosts, which do not distinguish between the plagiarized calls and those uttered by their own kind.

Hen viduines only come into breeding fettle by watching the breeding preparations of the host species.

Protective Associations

Although the siting and camouflage of the nest, the eggs, the young and the incubating parent may help to evade danger, relatively defenceless species may achieve some protection by associating with more aggressive animals.

Colonial birds derive a considerable benefit by operating on the principle of safety in numbers and this topic will be dealt with later. Very often, a number of closely related species nest together as in auks, gulls and herons.

In South America, small tanagers and tyrant flycatchers tend to nest in the vicinity of aggressive birds like the Kiskadee, which shows no tolerance or fear of larger predatory species. Similarly, Red-tailed Buzzards, Marabou Storks, Snowy Owls and Pied Crows quite often have defenceless birds nesting around their own nest sites.

There are many fascinating examples of birds shielding behind aggressive insects, presumably for protection. Black-throated Warblers and South American Yellow-rumped Caciques nest around the pendant dwellings of vicious hornets and wasps. In Africa, in places where thorny acacia trees abound, weaverbirds nest in trees previously selected by wasps as sites for their conspicuous nests. Why these insects should tolerate their avian neighbours is quite puzzling. Some birds actually excavate their breeding places in the homes of ants, wasps or termites; among the more ambitious lodgers are Australian kingfishers, parrots from Brazil and Papua and a woodpecker. It seems almost inconceivable that their hosts should take so readily to their avian guests, particularly as, in some cases, the insects are taken as food.

Even man may unwittingly provide nesting places for birds that are comparatively free from predators. Various kinds of pigeons, martins, swifts, weavers, the ubiquitous House Sparrow and starlings now exploit sites offered by our buildings which, so far as the birds are concerned, are no different from their previous cliff haunts. Sometimes, the species are more unusual, such as White and Abdim's Storks, and even in the middle of Amsterdam there is a thriving heronry even though the birds have to fly some way in order to catch their food. Towns often harbour rookeries and the birds breed without trouble, despite the bustle below.

Several birds live in association with aggressive insects, probably for protection. In South America, Yellow-rumped Caciques build their hanging nests alongside pendant wasps' nests.

Nest Sanitation

Keeping the nest clean would appear to have good survival value because it would help to keep disease and parasites in check. In Passerines, egg shells and faecal sacs may be removed or eaten by the parents. When the young are older, the faeces are squirted beyond the rim of the nest, but in species like the Goldfinch, they remain encrusted around the nest cup.

The survival value of removing the egg shells has been demonstrated for the Black-headed Gull. As the inside of the egg is white, fragments stand out and would attract predators to the newly-hatched chicks. Egg shells are, therefore, carried away from the territory after a short time, an action that takes only a few seconds and yet the whole future of the nestlings may depend upon it. However, not all gulls show this be-

A parent Reed Warbler removing a faecal pellet produced by its young.

haviour. Herring Gulls are large and ferocious enough to take care of most predators and Kittiwakes nesting securely on cliff faces are relatively unworried by Carrion Crows.

Excrement may be used as a sophisticated form of camouflage. Most gulls and terns defaecate away from the nest site but Sandwich Terns seem to be an exception. These birds nest in compact groups often among colonies of other gulls or terns, and furthermore the eggs are relatively well camouflaged against surroundings 'whitewashed' by excreta. Recent investigations have shown that predators might tend to keep away from the 'whitewashed' tern areas and, if they do venture in, they would tend to miss the Sandwich Tern eggs because they would still be looking for those of the gulls which would stand out against the terns' excreta.

A Black-headed Gull removing pieces of shell from its nest. The white inside of the shell would show up against the dark nest, and attract the attention of predators such as Carrion Crows (*inset*).

Signalling for Food

The young of some birds that hatch in an advanced state of development, such as ducks and game birds, soon learn to feed themselves; however, the chicks of the majority of birds must depend entirely upon their parents to bring them nourishing food. Feeding is really a co-operative exercise between the parent and offspring; the parent has the food and will give it to the youngsters providing they respond correctly with the right kind of begging behaviour.

In the much-studied Herring Gull, the newborn chick encourages the parent to regurgitate food from its crop by pecking at the red spot on the parent's lower mandible, a habit that occurs in other gulls. Young that are born nearly naked and blind respond to any nest movement by opening their bills widely and either craning their necks vertically (in the case of open-cup nesters) or obliquely sideways over their shoulders, as in the grassfinches. Later on when the eyes become opened, the begging brood orientate visually towards the parent, wherever it may be.

The most characteristic feature of a young bird which depends upon its parent for food is often its gape; those of Passerine nestlings are usually greatly exaggerated and the insides of the mouths are brightly coloured. These factors act

An American Robin feeding its young

as potent signals that stimulate the parent to feed and, at the same time, act as guides or targets for the adult's bill. Sometimes these gape markings are highly elaborate as in the Bearded Reedling or grassfinches; colourful refractive globules are set in the mouth flanges and palate, and in the Horned Lark the tongue is adorned with jet-black spots and bars. These oral displays, although transitory, are very specific and the parents will only feed those mouths with the correct pattern of markings characteristic of their own species.

Some birds, like the waxbills, no doubt are able to recognize their own young by instinct; they do not have to learn and consequently it is not very easy to swop around chicks experimentally between different species. However, lovebirds learn the characteristics of their young during their first period of parenthood. If they are allowed to bring up chicks of another species during their first brood, then they will subsequently reject their own for the foster-chicks.

In many hole-nesting birds, vocalizations probably play a greater role than visual signals in encouraging the parents to bring food. If the hunger calls of several chicks are played to a pair of Pied Flycatchers with only one chick, then they will bring in prodigious quantities of insects, far more than necessary, for their satiated offspring.

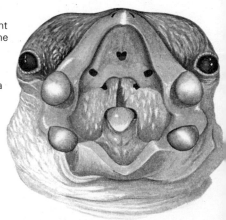

The pattern of markings on the inside of the mouth of a newly-hatched bird may be quite elaborate, as in an infant Parrot Finch (*right*). When the young bird opens its mouth, the resulting *gape display* stimulates the parent to feed the young, and also acts as a guide for the parent's bill.

Herring Gull

Laughing Gull

Lesser Black-backed Gull

Black-headed Gull

Young birds often fail to distinguish between crude representations of their parents and their real ones, and so models have been used to find out more precisely the stimuli to which the chicks respond.

Take the European Herring Gull. The adults have an orange spot on the lower mandible of the otherwise yellow bill; the newly-hatched chick pecks at it and the parent then regurgitates food. By presenting the chicks with different coloured models, it has been found that they were more sensitive to *red* than to other colours and would peck more persistently at *thin* objects, such as a red pencil, than at a model of a thick red bill. The exaggerated pecking response towards slender red stimuli may explain why the bills of the smaller gulls like the Black-headed Gull of Eurasia and Laughing Gull of North America are fully red – they may be around the optimum size to elicit this basic response. On the other hand, the larger Herring Gull and Greater and Lesser Black-backed Gulls which have deeper bills to deal with heavier food, have retained a thin red spot on their bills rather than evolve thick red bills which might be less effective in causing their young to beg.

To what extent are the stimuli to which the chick responds related to the feeding situation? Arctic Tern chicks peck at strips of red, blue and silver paper more than to strips of other colours. They will also solicit food from head profiles with red bills but these become more effective if a strip of silver paper is placed in the bill to simulate an adult bringing back a fish (*right*). The model that was found to be the most successful in arousing the chicks to beg consisted of a silver bill with a red 'fish', so the colours need not be orientated the same way as in the natural feeding situation of Arctic Terns.

Family Protection

Recognition of the Young in Zebra Finches

In species where the cocks are highly coloured, the juveniles often take after the hens, although there may be subtle differences and the features that distinguish the young act as signals for inducing parental responses. In Zebra Finches, the adults have scarlet bills whereas their offspring have black bills for the first few weeks of life. If the fledglings are taken and their bills painted red, their father will mistake them for adult hens and attempt to court them. Also, young with black bills are fed in preference to those with their bills painted red although both may beg just as furiously. Parental responses then might be influenced by very simple features.

It is no accident that immature birds often have a coloration

Adult hen
Zebra Finch

Young Zebra
Finch

The hen will ignore a young bird whose black bill has been painted red.

fairly different from that of their parents. Young birds must be able to broadcast the fact that they are not likely to compete with their elders for commodities that they demand, such as mates and territories. Immature Gannets, for example, far from being the dazzling white of their parents, are dark grey. The Kittiwake's black collar emphasizes an appeasement ceremony (head flagging) and therefore helps directly to cut down adult hostility. Sometimes, the plumage coloration complements special juvenile behaviour; for example, when danger threatens many young gulls and terns crouch and have evolved camouflaged plumage unlike their conspicuous parents who can flee. Young gulls may scavenge along the shore more than their elders, and in this habitat their darker plumage may conceal them more effectively than the pale adults.

Young Robin

Adult Robin

Young Robins lack red breast feathers that would provoke adults to instant attack. The young Night Heron (*far right*) has a camouflaged plumage most unlike that of its parent.

Family Protection in Seabirds

The way in which nests can be sited so that they receive the maximum protection has already been discussed briefly. However, there is often a clear relationship between the kind of security offered by the nest site and that given by the parent's behaviour. Nest siting, reactions of the parents and camouflaging all affect the survival of the young.

Clearly, species that produce initially helpless young are prone to the greatest dangers and these birds often nest in quite inaccessible places, such as cliff ledges in the case of auks. Large numbers assemble in traditional sites where the young will be relatively safe. Concentrations of birds are likely to attract predators, and in these birds, the nature of the nest sites is all the protection that the young auks have.

Gulls and terns often nest on islets or shingle spits that are only partially protected from land predators such as foxes and avian ones like Carrion Crows. The chicks can move around the area in the immediate vicinity of the nest. Most gulls are basically white and very conspicuous; they would easily draw the attention of marauders and so, at the first sign of danger, they *leave* the well-camouflaged eggs or chicks. Alarm cries cause the young to hide and *crouch* – an important response because should they wander from their parent's territory they are likely to be killed by the neighbours. Colonial nesting is a means of combating predators, which are mercilessly mobbed, harried and even attacked by the adults whose territory they violate. The nests must not be too close together as this would help foxes, but they must not be too widely dispersed because the effectiveness of the mobbing would decrease.

The Golden Plover's
broken-wing trick

Some species are particularly vigorous in the defence of their nest sites and young; many owls, terns and skuas may draw blood from humans who venture too close. More subtle forms of defence are chiefly used by birds that nest in open habitats; these can be called distraction displays and generally fall into two categories. Injury feigning is perhaps the best known of these. Calling plaintively, plovers draw the attention of an intruder by running along the ground dragging a wing as though it were broken; the marauder thinks that the bird is injured, it follows and suddenly the 'lame' bird takes flight.

Another highly-ritualized distraction display is performed by waders such as the Purple Sandpiper that breed in areas where the main terrestrial predator is the Actic Fox. This mammal preys chiefly upon rodents and so should one threaten the brood, the parent birds run along the ground in a hunched posture with mincing steps looking like a rodent.

The function of these displays is to draw the attention of predators from the vulnerable brood and they are performed when the impulse to escape conflicts with the desire to stay and defend the young.

Nursery Groups or Crèches

At some stage during their infant life, the young of Eider, flamingos and several of the more southerly species of penguins gather into crèches. In penguins these groups are protective and operate on the principle of 'safety in numbers'. Skuas are among the penguin chicks' chief predators but, bold though these birds are, they will not enter the crèche. Chicks only join the groups when both parents have to go off together to collect food and they leave it when they are large enough to fend for themselves.

LEARNING WHO'S WHO

We perhaps take it too much for granted that animals tend to sort themselves out neatly into their own kinds and rarely interbreed with others under natural conditions. Do birds recognize their own kind naturally or is the process influenced by experience? The answer might well vary from one species to another. For example, brood parasites never see their real parents and yet they choose correctly in selecting a mate later in life. There is, however, much evidence that many birds may have to learn the characteristics of their own species and that they do so during critical periods of their lives.

Imprinting

On hatching, ducklings and goslings generally follow the parent or parents who lead them away from the nest to open water. So strong is this following response that the young birds will attach themselves to any large moving objects or human being if they are deprived of their natural parents. At this time, the ducklings and goslings are especially receptive to the characteristics of the 'parent' whether this be the true parent, foster parent or human and they will in the future choose to make positive social, and maybe sexual, responses to individuals bearing these characteristics later on in life. This learning process is irreversible and is called *imprinting*.

An Eiderduck leading her brood to open water. At this time the young are rapidly absorbing the characteristics of their own species and becoming 'socialized'.

144

There is a critical age when young birds are most strongly imprinted and in the Mallard duck it reaches a height between the thirteenth and sixteenth hour of life; by the twenty-ninth hour the strength of the imprinting is reduced almost to zero. So ducklings forced to accept strange substitutes as mothers or foster parents during the first two days of life may never learn to live with individuals of their own kind, but prefer those of the foster parent type as social companions.

The strength of imprinting depends upon the effort exerted by the duckling in following its parent. This was shown in an experiment in which ducklings were made to clamber over hurdles in order to keep up with a model. These ducklings showed a firmer attachment to their 'parent' than others that had had an easier time.

Imprinting may be strong in goslings. If they are deprived of proper parents soon after birth, goslings will accept human beings as 'mothers'. Later on, these birds will prefer humans for company rather than other geese.

There is no reason why the concept of imprinting should be restricted to visual learning; in the hole-nesting Wood Duck, the ducklings become imprinted to the calls of the parent during the first few hours after hatching. In this case the parent has to entice them out of the dark nest by calling.

The phenomenon discussed above also occurs in rails, Moorhens and Coots.

Young ducklings will clamber over hurdles to keep up with models of their parents.

In a sense birds tend to look for reminders of their parents (natural or otherwise) in their mates. For example, domesticated pigeons come in all colours and a long time ago it was realized that they preferentially choose mates that resemble their parents in coloration. Mallard drakes brought up by a duck Pintail become imprinted on her and will later on force their attentions on duck Pintails. This does not necessarily work the other way round because ducks have an *inborn* preference for the drake of their own species.

Even small grassfinches go through a sensitive period when they learn the special features of their own species, and this occurs between the thirtieth and fortieth day after hatching. Bengalese Finches, a highly domesticated form of the Asian Striated Finch, are often used by aviculturists as foster parents; experience shows that the birds of other species that they rear so well often fail to mate themselves. Cock Zebra Finches brought up by Bengalese Finches will always prefer to mix with and court members of their foster parent species even if given the opportunity of mating with their own kind. Experiments

A Mallard drake courting a hen Pintail

performed by Dr Klaus Immelmann in Germany have shown that just after the thirtieth day of life the features of the 'parents' become indelibly stamped in the juvenile bird's brain and this governs its choice of companions and mates for the rest of its life.

In those species in which the parents are quite different, the young cocks may have to learn which of its parental types should be courted. In Zebra Finches, it is the cocks that chase away the brood between the thirty-fifth and fortieth day and the hostility of the father, which is brightly coloured and quite different from the drab mother, somehow ensures that his sons will not direct their courting energies towards other cocks. Young cock Zebra Finches isolated from their parents before the thirty-fifth day court both cocks and hens when they become sexually mature.

If young cock Zebra Finches are brought up by a hen Bengalese Finch (*right*), they will court hen Bengalese Finches when adult (*below*) even if they have the opportunity of mating with their own kind.

MIGRATION

For thousands of years people have watched the great annual movements of birds and wondered about the mysterious driving force behind them. Now we know that whole populations of certain species shift northwards and southwards with the seasons. Giant Petrels and many of the albatrosses, however, that nest on isolated oceanic islands circumnavigate the world on the Roaring Forties and other trade winds between their breeding seasons. It is not difficult to guess that migration has survival value as migrating species tend to follow favourable weather. Insect eaters take advantage of the bloom of food in northern latitudes in spring and summer, rear their young and then seek better areas when the weather turns harsher and makes their food scarce.

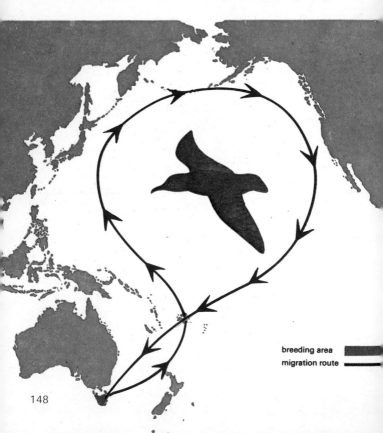

breeding area
migration route

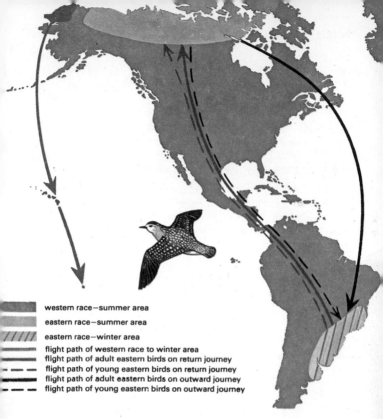

western race—summer area
eastern race—summer area
////// eastern race—winter area
flight path of western race to winter area
flight path of adult eastern birds on return journey
- - - flight path of young eastern birds on return journey
flight path of adult eastern birds on outward journey
- - - flight path of young eastern birds on outward journey

Migration routes of the Slender-billed Shearwater (*opposite*) and the Pacific Golden Plover (*above*).

Long Distance Travellers

It has been said that the length of birds' migrations is limited by the size of this planet! Arctic Terns may hold the record; they breed as far north as latitude 82°N and afterwards travel down the Atlantic Coast of Europe and Africa to the Antarctic seas at 74°S, a round trip of about 36,000 miles. Terrestrial birds may also make prodigious journeys over the sea; the Pacific Golden Plover which nests in Alaska and Siberia wings its way 2,000 miles across the Pacific Ocean to Hawaii. Even the diminutive Ruby-throated Hummingbird buzzes non-stop for 500 to 1,000 miles across the Gulf of Mexico.

149

Tracking Migrants

The sight of vast flocks of birds making their way purpose-fully across the skies is a thrilling spectacle, and much of our knowledge of these movements, such as their timing, relation-ship to the weather and the species taking part has been built up by direct observations at bird observatories.

Banding or ringing has been an invaluable technique, particularly now that many countries have their own schemes. In the British Isles alone, over $5\frac{1}{4}$ million birds were ringed between 1909 and 1966, and 144,279 were recovered. By analysing these recoveries, it is possible to build up a picture of the dispersal patterns for each species.

The widespread use of high-powered radar equipment, which shows up birds or flocks as spots or patches of light on

Recoveries throughout Europe and North Africa of Wheaters ringed in the British Isles show the migration routes taken by these birds.

the Plan Position Indicator (PPI), has revealed that much migration takes place at great heights. Over England the most frequent altitude seems to be around 2,700 to 3,000 feet although a considerable amount of avian traffic occurs well above this height. Radar can be used for monitoring the routes taken by flocks over wide areas and also for detecting nocturnal migrants; indeed it seems as though much bird migration takes place under the cover of darkness when there is little danger from predators.

Radar has even cast some doubt upon the validity of visible migration watches because very often it seems as though the main-stream high-altitude bird movements bear little relation to what ornithologists report from the ground; these low-level birds are, in fact, 'lost' ones that have become disorientated.

Ringing has shown that the Irish and Scottish populations of Barnacle Geese have different breeding areas.

Spitsbergen

Greenland

to Siberia

Solway
Firth

151

Living Compasses

Migration behaviour is triggered off by a variety of circumstances, but the chief one is probably day length. The ever-shortening days at the end of the breeding season induces a change in the hormonal balance; this change causes the birds to feed avidly and lay up reserve fuel for their impending journey in the form of layers of fat. They also become increasingly restless. This activity generally occurs at night when migration flights usually start and if birds like white-throats or starlings are kept in cages at this time, they show an ability to select a compass bearing by attempting to fly in a particular direction, e.g. south.

Migrating birds must have the ability to navigate, although this does not imply that they know where they are heading,

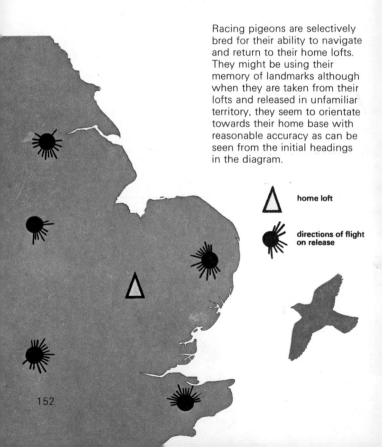

Racing pigeons are selectively bred for their ability to navigate and return to their home lofts. They might be using their memory of landmarks although when they are taken from their lofts and released in unfamiliar territory, they seem to orientate towards their home base with reasonable accuracy as can be seen from the initial headings in the diagram.

△ home loft

☀ directions of flight on release

and this has been demonstrated in many species by giving them the opportunity to 'home'.

All the evidence points to the fact that birds use the sun and stars as navigation beacons in the same way as we do, because migrating birds become hopelessly lost when the sky is overcast. It must be assumed that each individual of a migratory species has an inborn capacity to fix its position by and follow certain star patterns and the sun. Furthermore, the ability to compensate for the change of position of the heavenly bodies in the sky over the course of twenty-four hours would also be essential. Birds do have a very precise sense of timing (an internal biological clock) and this timing would help them to keep to a true heading with reference to celestial features. Without this there would be complete confusion.

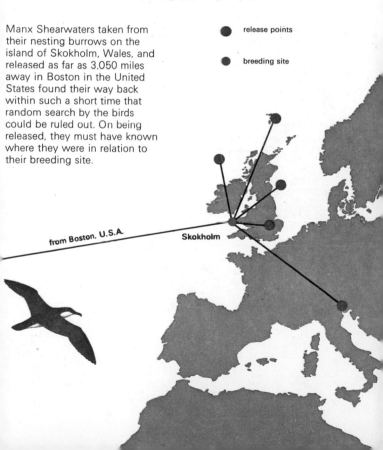

Manx Shearwaters taken from their nesting burrows on the island of Skokholm, Wales, and released as far as 3,050 miles away in Boston in the United States found their way back within such a short time that random search by the birds could be ruled out. On being released, they must have known where they were in relation to their breeding site.

release points

breeding site

from Boston, U.S.A.

Skokholm

The fact that migrating birds often set course on a fixed direction is illustrated by the following example. While migrating through Rossitten or Rybachi, a bird sanctuary near Kaliningrad in Russia, during spring, a large number of Hooded Crows were caught and transported westwards by 750 kilometres (470 miles). Of the birds recaptured, many were well displaced from their normal summer ranges and, therefore, had failed to compensate for the initial displacement (*opposite*).

Similar experiments have been performed with Starlings and, although young birds show a preference to take up a certain compass direction and keep to it, adults that have already migrated show an ability to compensate for artificial displacement and turn up in their traditional winter or breeding grounds.

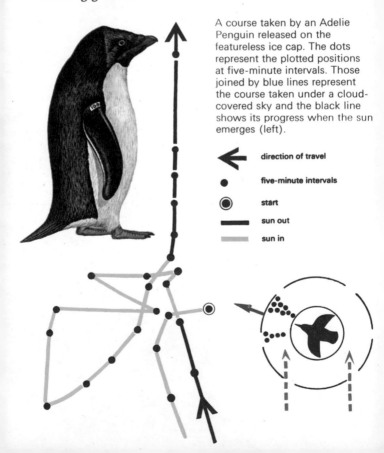

A course taken by an Adelie Penguin released on the featureless ice cap. The dots represent the plotted positions at five-minute intervals. Those joined by blue lines represent the course taken under a cloud-covered sky and the black line shows its progress when the sun emerges (left).

direction of travel

five-minute intervals

start

sun out

sun in

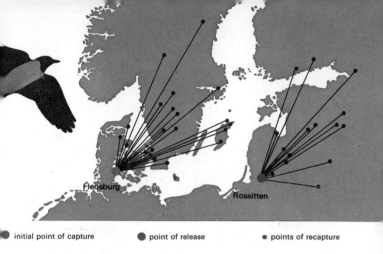

● initial point of capture	● point of release	● points of recapture

Having an inbuilt sun-and-star compass orientation is not enough to account for the amazing journeys that birds undertake. They would be blown off-course by side winds, for example, and therefore compensation for drift must take place; the amount of drift may be estimated by watching landmarks, islands or stationary cloud formations over land.

So far as migration and navigation is concerned, we are just beginning to fill in a few details; there is a great deal yet to be discovered which may well alter the theories we already have.

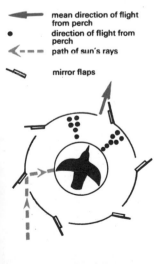

➡ mean direction of flight from perch
● direction of flight from perch
⬅--- path of sun's rays

⬋ mirror flaps

Directions in which Starlings attempt to fly in a specially-constructed cage when the apparent position of the sun is changed by mirrors. The birds were in a migrating mood (opposite).

BOOKS TO READ

General Books

Animal Behaviour by J. D. Carthy. Aldus, London, 1965.

Animal Behaviour by N. Tinbergen. Life Nature Library, 1966.

Birds of the World by O. L. Austin and A. Singer. Paul Hamlyn London, 1961.

Instructions to Young Ornithologists by D. Goodwin; Volume II, Birds Behaviour. Museum Press, London, 1961.

King Solomon's Ring by K. Lorenz. Methuen, London, 1961.

The Pictorial Encyclopedia of Birds by J. Hanzak, edited by Bruce Campbell. Paul Hamlyn, London, 1967.

The Science of Animal Behaviour by P. L. Broadhurst. Penguin, London, 1963 (Pelican series).

Social Behaviour in Animals by N. Tinbergen (second edition). Methuen, London, 1965.

The World of Birds by J. Fisher and R. T. Peterson. Macdonald, London, 1964.

What Bird is That? A Guide to the Birds of Australia by N. W. Cayley Angus and Robertson, Sydney and London, 1966 (fourth edition).

More Specialized Books

Courtship: a Zoological Study by M. Bastock. Heinemann, London, 1967.

Handbook of Waterfowl Behaviour by P. A. Johnsgard. Constable, London, 1966.

The Herring Gull's World by N. Tinbergen. Collins, London, 1953.

'Instinct' and 'Intelligence' by S. A. Barnett. MacGibbon and Kee, London, 1967.

The Life of Birds by J. C. Welty. Saunders, London, 1962.

The Life of the Robin by D. Lack. Witherby, London, 1965 (fourth edition).

Mechanisms of Animal Behaviour by P. R. Marler and W. J. Hamilton. Wiley, Chichester, 1966.

Penguins by J. H. Sparks and T. Soper. David and Charles, Newton Abbot, 1967.

Radar Ornithology by E. Eastwood. Methuen, London, 1967.

Social Behaviour from Fish to Man by W. Etkin et al. University of Chicago Press, 1967.

A Study of Bird Song by E. A. Armstrong. Oxford University Press, London, 1963.

INDEX

Page numbers in bold type refer to illustrations.

OTHER TITLES
IN THE SERIES